Converter and Filter Circuits

Other Books in this Series

Converter and Filter Circuits

Rudolf F. Graf

Newnes
Boston Oxford Johannesburg Melbourne New Delhi Singapore

Newnes is an imprint of Butterworth-Heinemann

Copyright © 1997 by Butterworth-Heinemann
Copyright © 1993 by Rudolf F. Graf

ℛ A member of the Reed Elsevier group

Library of Congress Cataloging-in-Publication Data
Graf, Rudolf F.
    Converter and filter circuits / Rudolf F. Graf.
        p.   cm.
    Originally published: The modern converter and filter circuit
encyclopedia. Bule Ridge Summit, PA : TAB Books, c1993.
    Includes index.
    ISBN 0-7506-9878-0
    1. Electric current converters. 2. Electric filters.
3. Frequency changers. I. Title.
TK7872.C8G68   1996
621.31'7—dc20   96-36496
                                       CIP

British Library Cataloguing-in-Publication Data
A catalogue record for this book is available from the British Library.

The publisher offers special discounts on bulk orders of this book.
For information, please contact:
Manager of Special Sales
Butterworth–Heinemann
313 Washington Street
Newton, MA 02158–1626
Tel: 617-928-2500
Fax: 617-928-2620

For information on all Newnes electronics publications available, contact
our World Wide Web home page at: http://www.bh.com/bh

Printed in the United States of America
10 9 8 7 6 5 4 3 2 1

Contents

Introduction

Like the other volumes in this series, this book contains a wealth of ready-to-use circuits that serve the needs of the engineer, technician, student and, of course, the browser. These unique books contain more practical, ready-to-use circuits that are focused on a specific field of interest than can be found anywhere in a single volume.

1

Analog-to-Digital Converters

The sources of the following circuits are contained in the Sources section, which begins on page 175. The figure number in the box of each circuit correlates to the source entry in the Sources section.

16-BIT A/D CONVERTER

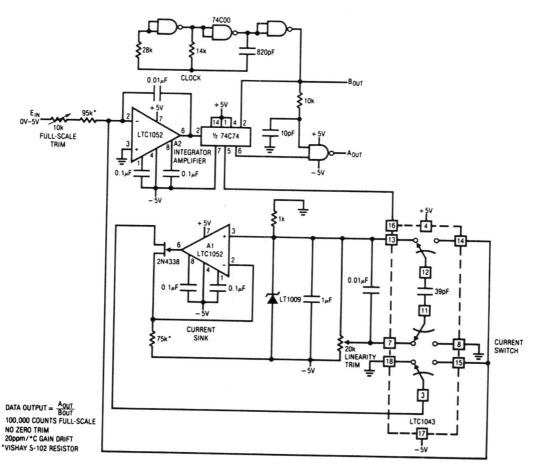

LINEAR TECHNOLOGY

Fig. 1-1

The A/D converter, made up of flip-flop A2, some gates, and a current sink, is based on a current-balancing technique. The chopper-stabilized LTC1052's 50 nV/°C input drift is required to eliminate offset errors in the A/D.

8-BIT A/D CONVERTER I

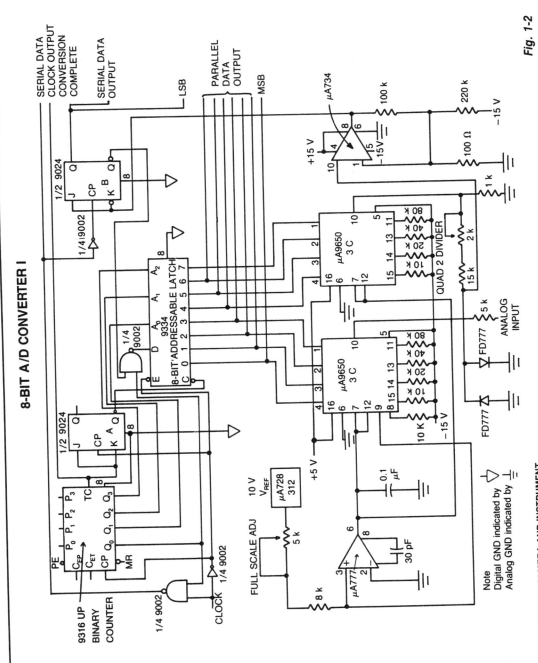

Fig. 1-2

FAIRCHILD CAMERA AND INSTRUMENT

HIGH-SPEED 12-BIT A/D CONVERTER

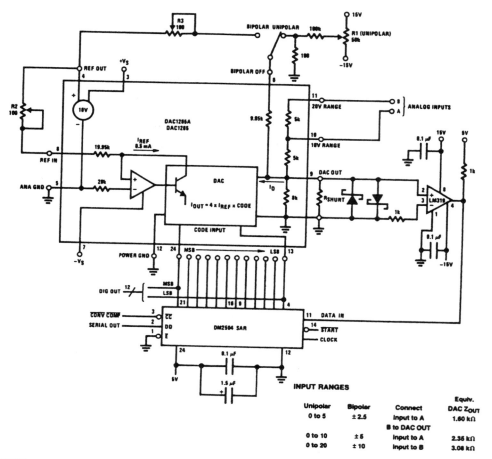

Unipolar	Bipolar	Connect	Equiv. DAC Z_{OUT}
0 to 5	±2.5	Input to A	1.60 kΩ
		B to DAC OUT	
0 to 10	±5	Input to A	2.36 kΩ
0 to 20	±10	Input to B	3.06 kΩ

INPUT RANGES

NATIONAL SEMICONDUCTOR

Fig. 1-3

This system completes a full 12-bit conversion in 10 μs unipolar or bipolar. This converter is accurate to ±1/2 LSB of 12 bits and has a typical gain TC of 10 ppm/°C. In the unipolar mode, the system range is 0 V to 9.9976 V, with each bit having a value of 2.44 mV. For true conversion accuracy, an A/D converter should be trimmed so that given bit-code output results from input levels from 1/2 LSB below to 1/2 LSB above the exact voltage, which that code represents. Therefore, the converter zero point should be trimmed with an input voltage of 1.22 mV; trim R1 until the LSB just begins to appear in the output code (all other bits "0"). For full-scale, use an input voltage of 9.9963 V (10 V-1 LSB-1/2 LSB); then trim R2 until the LSB just begins to appear (all other bits "1"). The bipolar signal range is −5.0 to 4.9976 V. Bipolar offset trimming is done by applying a −4.9988-V input signal and trimming R3 for the LSB transition (all other bits "0"). Full-scale is set by applying 4.9963 V and trimming R2 for the LSB transition (all other bits "1").

4

SUCCESSIVE-APPROXIMATION A/D CONVERTER I

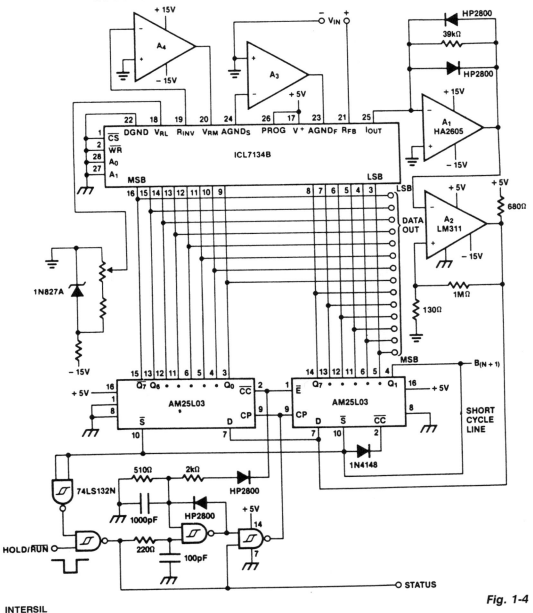

INTERSIL

Fig. 1-4

A bipolar input, high-speed A/D converter uses two AM25L03s to form 14-bit successive approximation register. The comparator is a two-stage circuit with an HA2605 front-end amplifier used to reduce settling time problems at the summing node. Careful offset-nulling of this amplifier is needed.

8-BIT A/D CONVERTER II

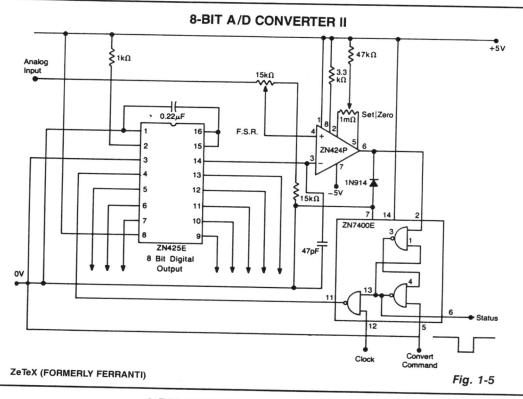

ZeTeX (FORMERLY FERRANTI)

Fig. 1-5

8-BIT TRACKING A/D CONVERTER

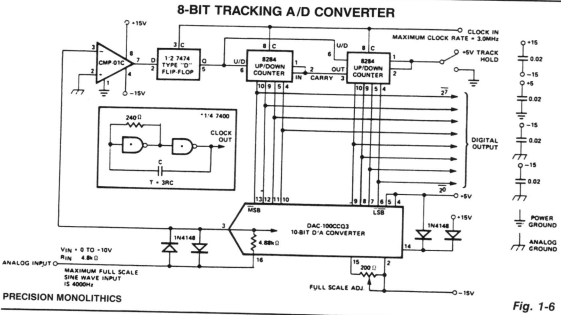

PRECISION MONOLITHICS

Fig. 1-6

SUCCESSIVE-APPROXIMATION A/D CONVERTER II

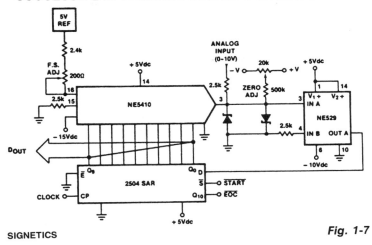

SIGNETICS

Fig. 1-7

The 10-bit conversion time is 3.3 μs with a 3-MHz clock. This converter uses a 2504 12-bit successive-approximation register in the short-cycle operating mode, where the end of conversion signal is taken from the first unused bit of the SAR (Q_{10}).

CYCLIC A/D CONVERTER

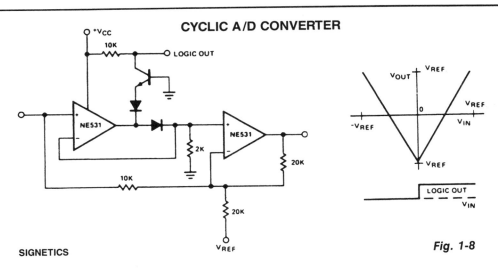

SIGNETICS

Fig. 1-8

The cyclic converter consists of a chain of identical stages, each of which senses the polarity of the input. The stage then subtracts V_{REF} from the input and doubles the remainder if the polarity was correct. The signal is full-wave rectified and the remainder of $V_{IN} - V_{REF}$ is doubled. A chain of these stages gives the gray code equivalent of the input voltage in digitized form related to the magnitude of V_{REF}. Possessing high potential accuracy, the circuit using NE531 devices settles in 5 μs.

TRACKING ADC

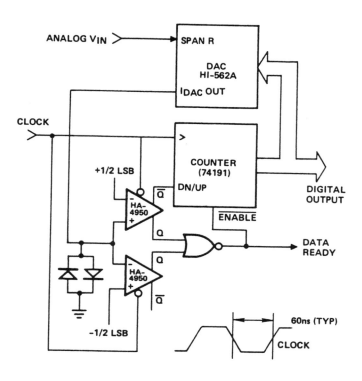

HARRIS

Fig. 1-9

The analog input is fed into the span resistor of a DAC. The analog input voltage range is selectable in the same way as the output voltage range of the DAC. The net current flow through the ladder termination resistance (i.e., 2 kΩ for HI-562A) produces an error voltage at the DAC output. This error voltage is compared with $1/2$ LSB by a comparator. When the error voltage is within $\pm 1/2$ LSB range, the Q output of the comparators are both low, which stops the counter and gives a data-ready signal to indicate that the digital output is correct. If the error exceeds the $\pm 1/2$-LSB-range, the counter is enabled and driven in an up or down direction, depending on the polarity of the error voltage.

The digital output changes state only when there is a significant change in the analog input. When monitoring a slowly varying input, it is necessary to read the digital output only after a change has taken place. The data ready signal could be used to trigger a flip-flop to indicate the condition and reset it after readout. The main disadvantage of the tracking ADC is the time required to initially acquire a signal; for a 12-bit ADC, it could be up to 4096 clock periods. The input signal usually must be filtered so that its rate of change does not exceed the tracking range of the ADC—1 LSB per clock period.

INEXPENSIVE, FAST 10-BIT SERIAL OUTPUT A/D

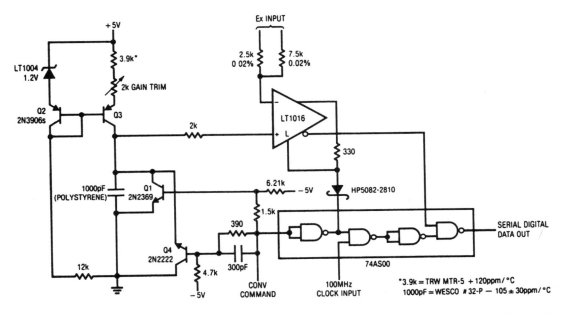

LINEAR TECHNOLOGY

Fig. 1-10

Every time a pulse is applied to the convert command input, Q1 resets the 1000-pF capacitor to 0 V. This resetting action takes 200 ns of the falling edge of the convert-command pulse, and the capacitor begins to charge linearly. In precisely 10 μs, it charges to 2.5 V. The 10-μs ramp is applied to the LT1016's positive input. The LT1016 compares the ramp to Ex, the unknown, at its negative input. For a 0- to 2.5-V range, Ex is applied to the 2.5-kΩ resistor. From a 0- to 10-V range, the 2.5-kΩ resistor is grounded and Ex is applied to the 7.5-kΩ resistor. Output of the LT1016 is a pulse, whose width directly depends on the value of Ex. This pulse width is used to gate a 100-MHz clock. The 100-MHz clock pulse bursts that appear at the output are proportional to Ex. For a 0- to 10-V input, 1024 pulses appear at full-scale, 512 at 5.00 V, etc.

9

8-BIT SUCCESSIVE-APPROXIMATION A/D CONVERTER

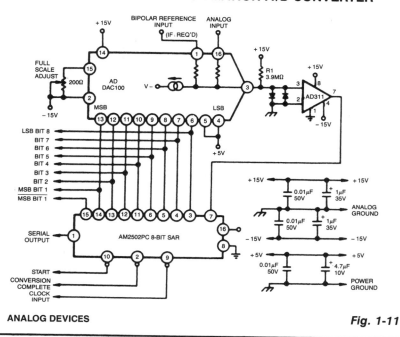

ANALOG DEVICES

Fig. 1-11

4-CHANNEL DIGITALLY MULTIPLEXED RAMP A/D CONVERTER

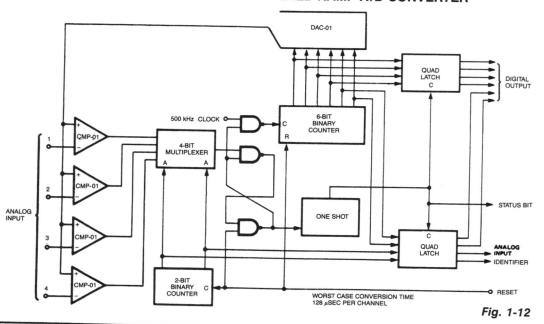

Fig. 1-12

ADC

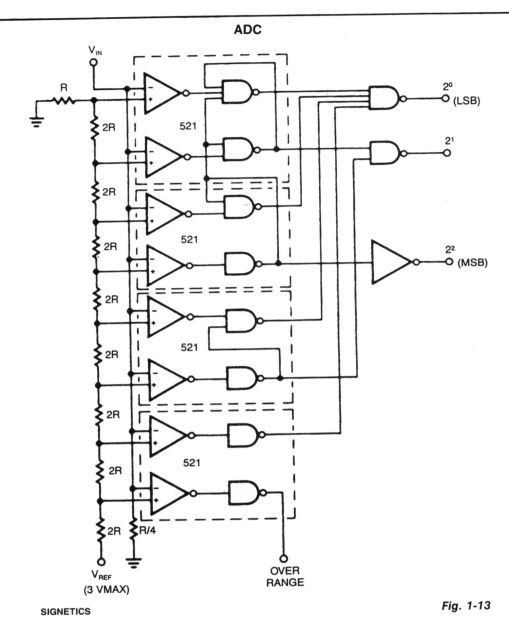

Fig. 1-13

Conversion speed of this design is the sum of the delay through the comparator and the decoding gates. Reference voltages for each bit are developed from a precision resistor ladder network. Values of R and $2R$ are chosen so that the threshold is $1/2$ of the least significant bit. This assures maximum accuracy of $\pm 1/2$ bit. The individual strobe line and duality features of the NE521 greatly reduced the cost and complexity of the design.

3-DECADE LOGARITHMIC A/D CONVERTER

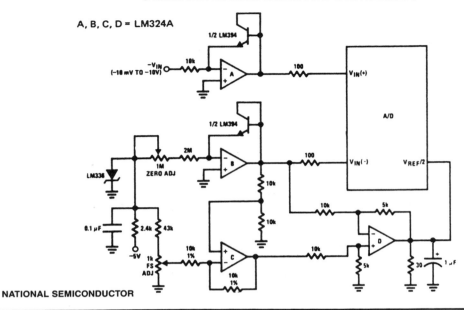

Fig. 1-14

TRACKING (SERVO TYPE) A/D CONVERTER

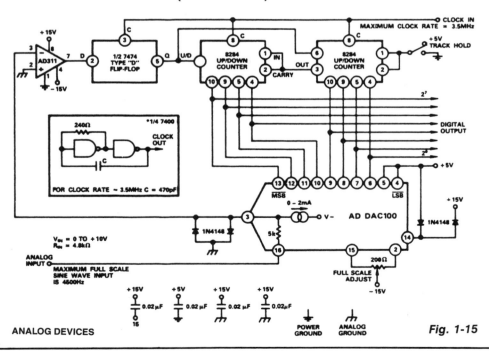

ANALOG DEVICES

Fig. 1-15

3¹/₂-DIGIT A/D CONVERTER WITH LCD DISPLAY

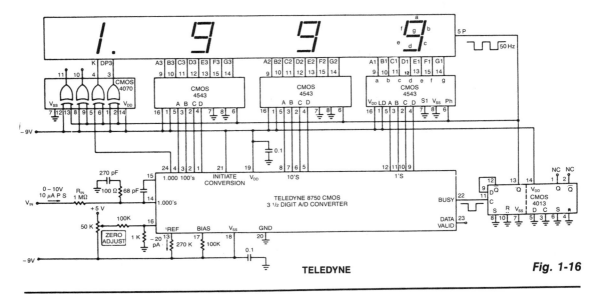

TELEDYNE

Fig. 1-16

FAST PRECISION A/D CONVERTER

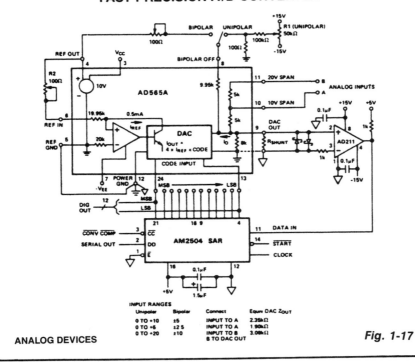

ANALOG DEVICES

Fig. 1-17

HIGH-SPEED 3-BIT A/D CONVERTER

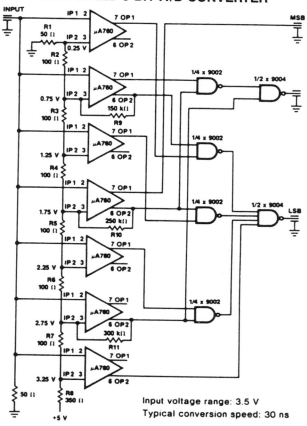

FAIRCHILD CAMERA
AND INSTRUMENT

Input voltage range: 3.5 V
Typical conversion speed: 30 ns

Fig. 1-18

3-IC LOW-COST A/D CONVERTER

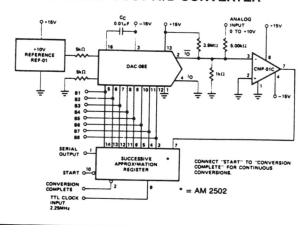

PRECISION MONOLITHICS

Fig. 1-19

SWITCHED-CAPACITOR ADC

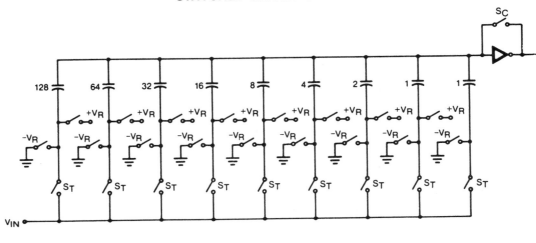

SILICONIX

Fig. 1-20

The CMOS comparator in the successive-approximation system determines each bit by examining the charge on a series of binary-weighted capacitors. In the first phase of the conversion process, the analog input is sampled by closing switch SC and all ST switches, and by simultaneously charging all the capacitors to the input voltage.

In the next phase of the conversion process, all ST and SC switches are opened and the comparator begins identifying bits by identifying the charge on each capacitor relative to the reference voltage. In the switching sequence, all 8 capacitors are examined separately until all 8 bits are identified, and then the charge-convert sequence is repeated. In the first step of the conversion phase, the comparator looks at the first capacitor (binary weight = 128). One pole of the capacitor is switched to the reference voltage, and the equivalent poles of all the other capacitors on the ladder are switched to ground. If the voltage at the summing node is greater than the trip point of the comparator—approximately 1/2 the reference voltage, a bit is placed in the output register, and the 128-weight capacitor is switched to ground. If the voltage at the summing node is less than the trip point of the comparator, this 128-weight capacitor remains connected to the reference input through the remainder of the capacitor-sampling (bit-counting) process. The process is repeated for the 64-weight capacitor, the 32-weight capacitor, and so forth down the line, until all bits are tested. With each step of the capacitor-sampling process, the initial charge is redistributed among the capacitors. The conversion process is successive-approximation, but relies on charge shifting rather than a successive-approximation register—and reference D/A—to count and weigh the bits from MSB to LSB.

SUCCESSIVE-APPROXIMATION A/D CONVERTER III

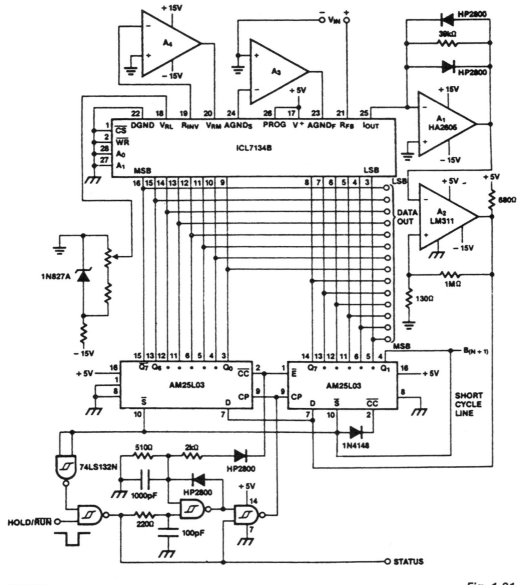

Fig. 1-21

SUCCESSIVE-APPROXIMATION A/D CONVERTER III (Cont.)

The ICL 7134B-based circuit is for a bipolar-input high-speed A/D converter, using two AM25L03s to form a 14-bit successive-approximation register. The comparator is a two-stage circuit with an HA2605 front-end amplifier, used to reduce settling time problems at the summing node. Careful offset-nulling of this amplifier is needed, and if wide temperature-range operation is desired, an auto-null circuit using an ICL7650 is probably advisable. The clock, using two Schmitt-trigger TTL gates, runs at a slower rate for the first 8 bits, where settling-time is most critical than for the last 6 bits. The short-cycle line is shown tied to the 15th bit; if fewer bits are required, it can be moved up accordingly. The circuit will free-run if the hold/run input is held low, but will stop after completing a conversion if the pin is high at that time. A low-going pulse will restart it. The status output indicates when the device is operating, and the falling edge indicates the availability of new data. A unipolar version can be constructed by typing the MSB (D13) on an ICL7134U to pin 14 on the first AM25L03, deleting the reference inversion amplifier A4, and tying V_{RFM} to V_{RFL}.

4-DIGIT (10,000 COUNT) A/D CONVERTER

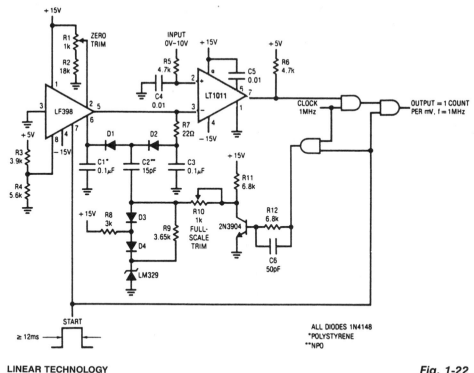

LINEAR TECHNOLOGY

Fig. 1-22

HALF-FLASH ADC

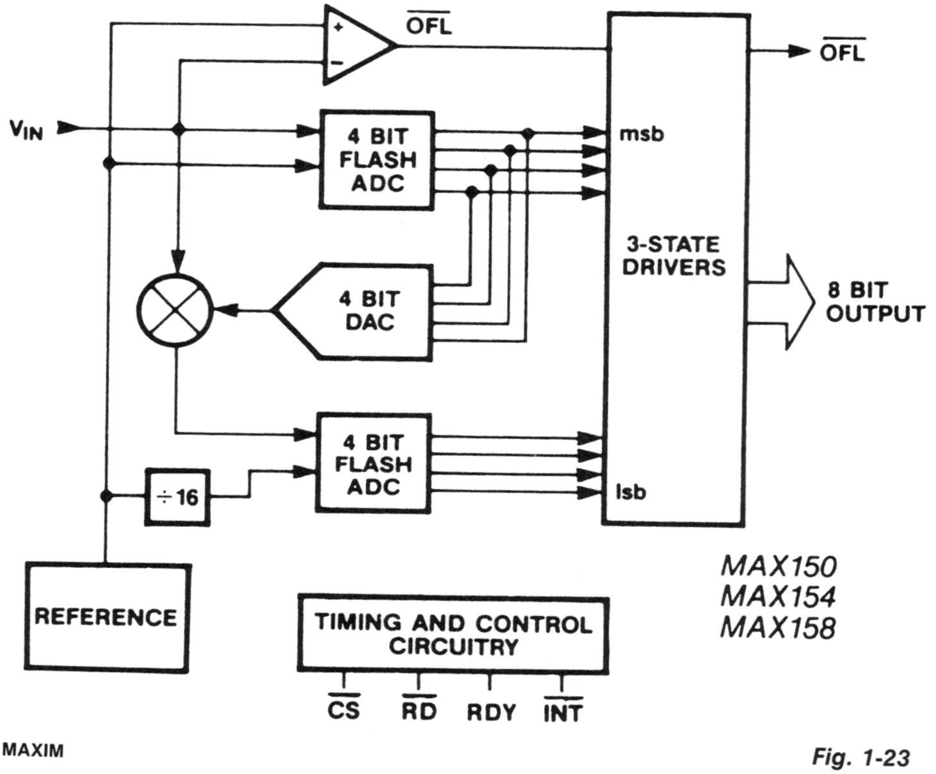

MAX150
MAX154
MAX158

MAXIM

Fig. 1-23

An A/D conversion technique that combines some of the speed advantages of flash conversion with the circuitry savings of successive approximation is termed *half-flash*. In an 8-bit, half-flash converter, two 4-bit flash A/D sections are combined. The upper flash A/D compares the input signal to the reference and generates the upper 4 data bits. This data goes to an internal DAC, whose output is subtracted from the analog input. Then, the difference can be measured by the second flash A/D, which provides the lower 4 data bits.

10-BIT A/D CONVERTER

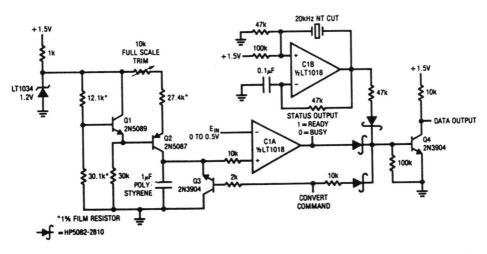

LINEAR TECHNOLOGY

Fig. 1-24

The converter has a 60-ms conversion time, consumes 460 μA from its 1.5-V supply and maintains 10-bit accuracy over a 15°C to 35°C temperature range. A pulse applied to the convert command line causes Q3, operating in inverted mode, to discharge through the 10-kΩ diode path, forcing its collector low. Q3's inverted mode switching results in a capacitor discharge within 1 mV of ground. During the time the ramps' value is below the input voltage, C1A's output is low. This allows pulses from C1B, a quartz-stabilized oscillator, to modulate Q4. Output data appears at Q4's collector. When the ramp crosses the input voltage's value, C1A's output goes high, biasing Q4, and ending the output data. The number of pulses at the output is directly proportional to the input voltage. To calibrate, apply 0.5 V to the input and trim the 10-kΩ potentiometer for exactly 1000 pulses out each time the convert command line is pulsed.

DIFFERENTIAL-INPUT A/D SYSTEM

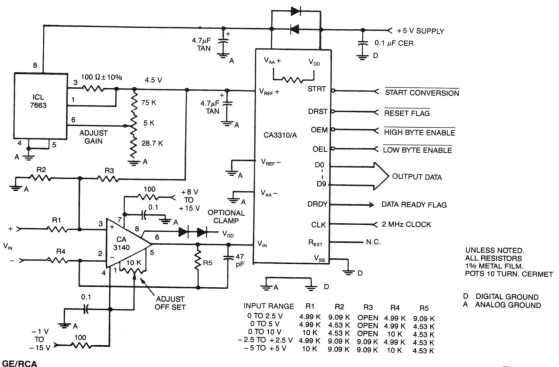

INPUT RANGE	R1	R2	R3	R4	R5
0 TO 2.5 V	4.99 K	9.09 K	OPEN	4.99 K	9.09 K
0 TO 5 V	4.99 K	4.53 K	OPEN	4.99 K	4.53 K
0 TO 10 V	10 K	4.53 K	OPEN	10 K	4.53 K
−2.5 TO +2.5 V	4.99 K	9.09 K	9.09 K	4.99 K	4.53 K
−5 TO +5 V	10 K	9.09 K	9.09 K	10 K	4.53 K

UNLESS NOTED.
ALL RESISTORS
1% METAL FILM.
POTS 10 TURN. CERMET

D DIGITAL GROUND
A ANALOG GROUND

GE/RCA

Fig. 1-25

Using a CA3140 BiMOS op amp provides good slewing capability for high-bandwidth input signals and can quickly settle energy that the CA3310 outputs at its V_{IN} terminal. The CA3140 can also drive close to the negative supply rail. If system supply sequencing or an unknown input voltage is likely to cause the op amp to drive above the V_{DD} supply, a diode clamp can be added from pin 8 of the op amp to the V_{DD} supply.

2

Current-to-Voltage Converters

The sources of the following circuits are contained in the Sources section, which begins on page 175. The figure number in the box of each circuit correlates to the source entry in the Sources section.

Current-to-Voltage Converter with Grounded Bias
   and Sensor
Current-to-Voltage Converter with 1% Accuracy
Current-to-Voltage Converter

CURRENT-TO-VOLTAGE CONVERTER WITH GROUNDED BIAS AND SENSOR

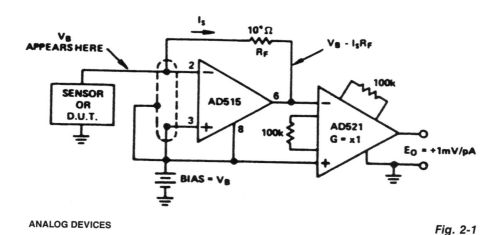

Fig. 2-1

CURRENT-TO-VOLTAGE CONVERTER WITH 1% ACCURACY

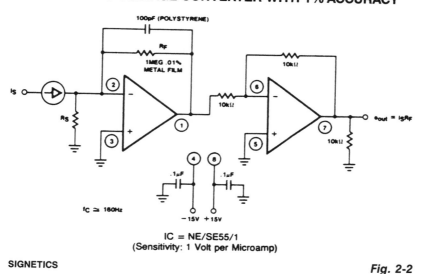

IC = NE/SE55/1
(Sensitivity: 1 Volt per Microamp)

Fig. 2-2

A filter removes the dc component of the rectified ac, which is then scaled to rms. The output is linear from 40 Hz to 10 kHz or higher.

CURRENT-TO-VOLTAGE CONVERTER

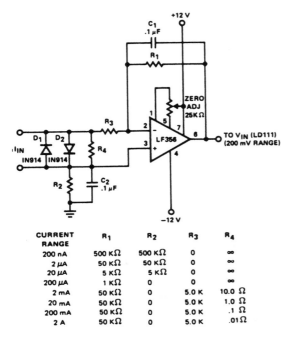

Converter features eight decades of current range. The circuit is intended to be used with the 200.0-mV range of a DVM.

CURRENT RANGE	R_1	R_2	R_3	R_4
200 nA	500 KΩ	500 KΩ	0	∞
2 µA	50 KΩ	50 KΩ	0	∞
20 µA	5 KΩ	5 KΩ	0	∞
200 µA	1 KΩ	0	0	∞
2 mA	50 KΩ	0	5.0 K	10.0 Ω
20 mA	50 KΩ	0	5.0 K	1.0 Ω
200 mA	50 KΩ	0	5.0 K	.1 Ω
2 A	50 KΩ	0	5.0 K	.01Ω

SILICONIX

Fig. 2-3

23

3

Digital-to-Analog Converters

The sources of the following circuits are contained in the Sources section, which begins on page 175. The figure number in the box of each circuit correlates to the source entry in the Sources section.

±10-V FULL-SCALE BIPOLAR DAC

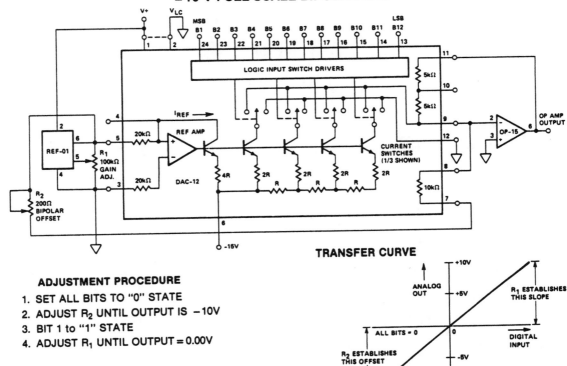

ADJUSTMENT PROCEDURE

1. SET ALL BITS TO "0" STATE
2. ADJUST R_2 UNTIL OUTPUT IS −10V
3. BIT 1 to "1" STATE
4. ADJUST R_1 UNTIL OUTPUT = 0.00V

TRANSFER CURVE

Fig. 3-1

PRECISION 12-BIT D/A CONVERTER

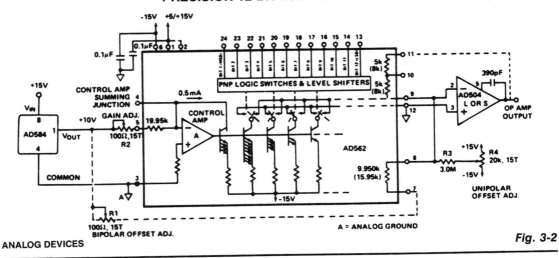

Fig. 3-2

12-BIT DAC

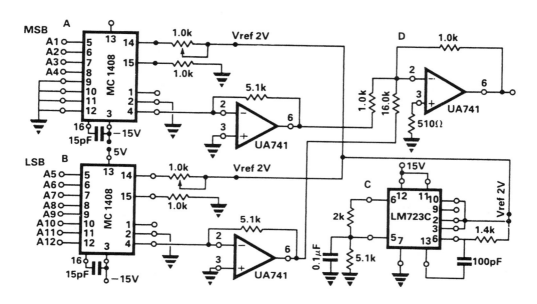

ELECTRONIC ENGINEERING

Fig. 3-3

Two MC1408s (8-bit D/A converters), A and B in the circuit diagram, are used. The four least-significant bits of A are tied to zero. The four most significant bits of the 12-bit data are connected to the remaining four input pins. The eight least-significant bits of the 12-bit data are connected to the eight input pins of B. The four most-significant bits of the 12-bit data together have a weight of 16, relative to the remaining eight bits. Hence, the output from B is reduced by a factor of 16 and summed with the output from A using the summing op-amp configuration D. Voltage regulator chip, LM7236, is used to provide an accurate reference voltage, 2 V, for the MC1408. The full-scale voltage of the converter is $1/6 \times 9.9609 + 1 \times (9.375) = 9.9976$ V. The step size of the converter is 2.4 mV.

RESISTOR-TERMINATED DAC (0- TO −5-V OUTPUT)

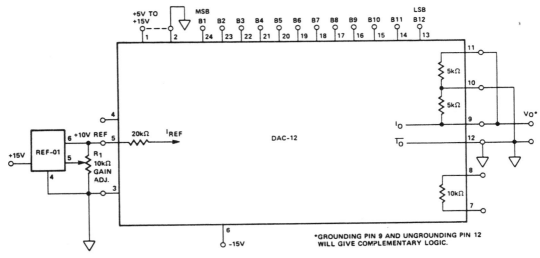

PRECISION MONOLITHICS

Fig. 3-4

3-DIGIT BCD D/A CONVERTER

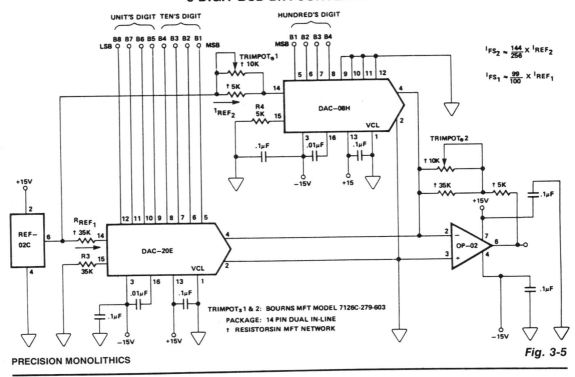

$$I_{FS_2} \approx \frac{144}{256} \times I_{REF_2}$$

$$I_{FS_1} \approx \frac{99}{100} \times I_{REF_1}$$

TRIMPOT₁ 1 & 2: BOURNS MFT MODEL 7126C-279-603
PACKAGE: 14 PIN DUAL IN-LINE
† RESISTORS IN MFT NETWORK

PRECISION MONOLITHICS

Fig. 3-5

±10-V FULL-SCALE UNIPOLAR DAC

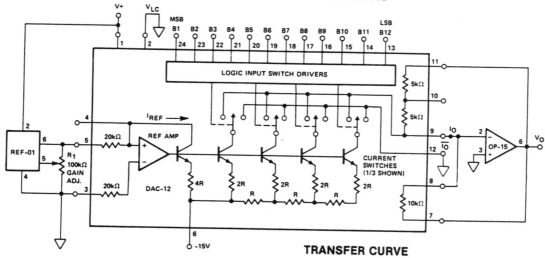

TRANSFER CURVE

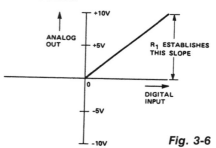

Fig. 3-6

ADJUSTMENT PROCEDURE

1. ALL BITS TO "1" STATE ("0" STATE IF PINS 9 AND 12 INTERCHANGED)

2. ADJUST R_1 UNTIL OUTPUT IS +9.9975

$$\frac{4095}{4096} \times 10V$$

PRECISION MONOLITHICS

HIGH-SPEED VOLTAGE-OUTPUT DAC

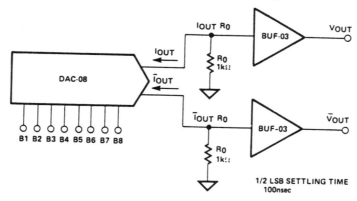

1/2 LSB SETTLING TIME 100nsec

SYSTEM WILL DRIVE CABLES OR TWISTED PAIRS.

PRECISION MONOLITHICS

Fig. 3-7

12-BIT BINARY 2's COMPLEMENT D/A CONVERSION SYSTEM

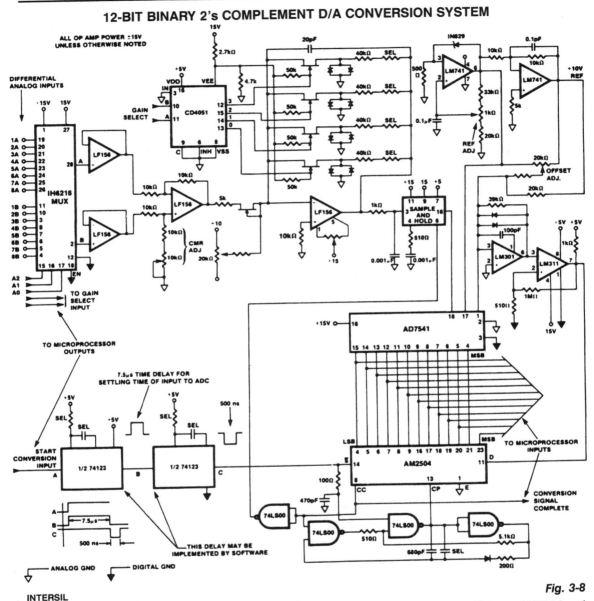

Fig. 3-8

INTERSIL

The front end of the DAC is configured differentially using dual 8-input IC multiplexer 1H6216 and three LM156 op amps. Following the differential amplifier is the programmable gain stage discussed earlier, with a low-pass filter on the output feeding the IH5110 sample-and-hold amplifier. The output of the IH5110 is connected to the comparator input, – input LM301, through the internal 10-kΩ feedback resistor of the 7541 multiplying D/A converter. The AD7541, along with a ±10-V reference and successive approximation logic, make up the 2's complement A/D converter.

12-BIT DAC WITH VARIABLE STEP SIZE

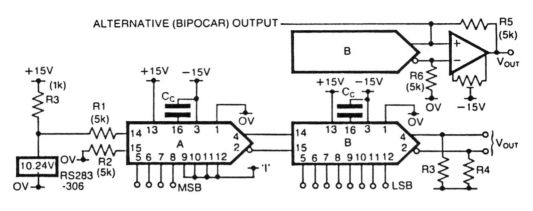

ELECTRONIC ENGINEERING

Fig. 3-9

The step size of the converter is variable by selection of the high-order data bits. The first DAC, A, has a stable reference current supplied via the 10.24-V reference IC and R1. R2 provides bias cancellation. As shown, only the first four MSB inputs are used, giving a step size of $225/256 \times 2.048/16 = 0.127$ mA. This current supplies the reference for DAC B, whose step size is then $0.1275/256 = 0.498$ μA. Complementary voltage outputs are available for unipolar output and using $R_3 = R_4 = 10$ kΩ, V_{out} is ± 10.2 V approximately, with a step size (1 LSB) of approximately 5 mV. If desired, an op amp can be added to the output to provide a low-impedance output with bipolar output symmetrical about ground, if $R_5 = R_6$ within 0.05%. Notice that offset null is required and all resistors, except for R2 and R3, should be 1% high-stability types.

By using lower order address lines than are illustrated for DAC A, a smaller step size (and therefore a full-scale output) can be obtained. Unused high-order bits can be manipulated high or low to change the relative position of the full-scale output.

9-BIT CMOS D/A CONVERTER

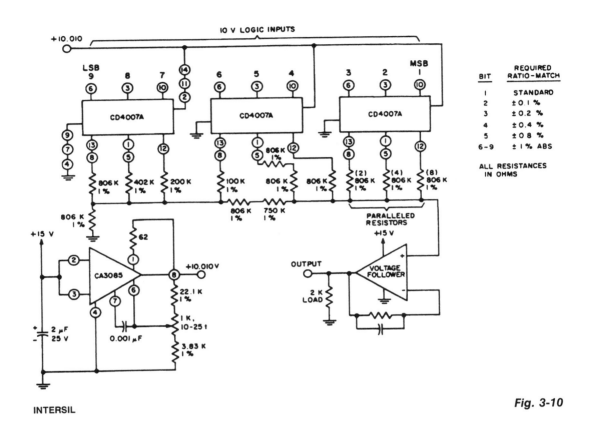

BIT	REQUIRED RATIO-MATCH
I	STANDARD
2	± 0.1 %
3	± 0.2 %
4	± 0.4 %
5	± 0 8 %
6 - 9	± 1 % ABS

ALL RESISTANCES IN OHMS

INTERSIL

Fig. 3-10

Three CD4007A IC packages perform the switch function using a 10-V logic level. A single 15-V supply provides a positive bus for the follower amplifier and feeds the CA3085 voltage regulator. The scale adjust function is provided by the regulator output control, which is set to a nominal 10 V in this system. The line-voltage regulation (approximately 0.2%) permits 9-bit accuracy to be maintained with a variation of several volts in the supply. System power consumption ranges between 70 and 200 mW; a major portion is dissipated in the load resistor and op amp. The regulated supply provides a maximum current of 440 μA of which 370 μA flows through the scale adjusting. The resistor ladder is composed of 1% tolerance metal-oxide film resistors. The ratio match between resistance values is in the order of 2%. The follower amplifier has the offset adjustment nulled at approximately a 1-V output level.

8-BIT D/A CONVERTER I

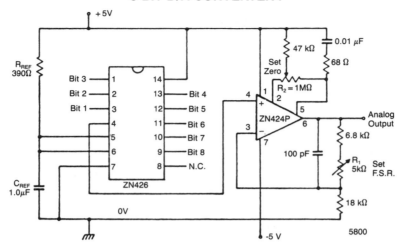

ZETEX (FORMERLY FERRANTI)

Fig. 3-11

HIGH-SPEED 8-BIT D/A CONVERTER

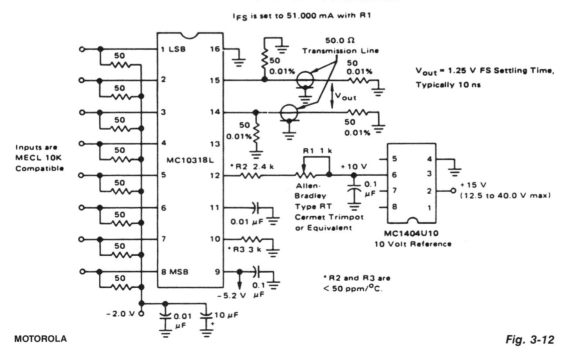

MOTOROLA

Fig. 3-12

14-BIT BINARY D/A CONVERTER (UNIPOLAR)

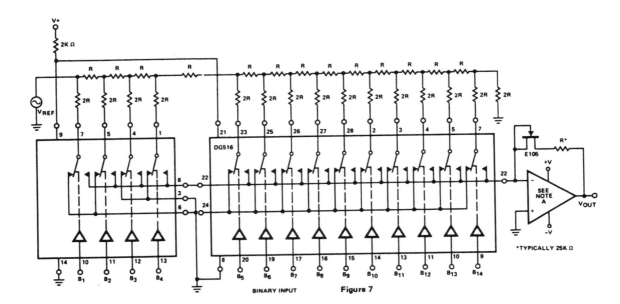

Figure 7

BINARY INPUT

NOTE:

A. Op-Amp characteristics effect D/A accuracy and settling time. The following Op-Amps, listed in order of increasing speed, are suggested:

   1. LM101A   2. LF156A   3. LM118

Unipolar Binary Operation

DIGITAL INPUT	ANALOG OUTPUT
1 1 1 1 1 1 1 1 1 1 1 1 1 1	$-V_{REF} (1 - 2^{-14})$
1 0 0 0 0 0 0 0 0 0 0 0 0 1	$-V_{REF} (1/2 + 2^{-14})$
1 0 0 0 0 0 0 0 0 0 0 0 0 0	$-V_{REF}/2$
0 1 1 1 1 1 1 1 1 1 1 1 1 1	$-V_{REF} (1/2 - 2^{-14})$
0 0 0 0 0 0 0 0 0 0 0 0 0 1	$-V_{REF} (2^{-14})$
0 0 0 0 0 0 0 0 0 0 0 0 0 0	0

SILICONIX

Fig. 3-13

10-BIT D/A CONVERTER

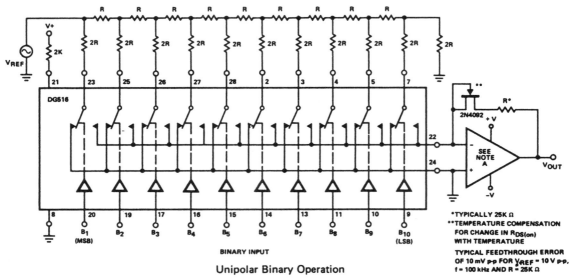

Unipolar Binary Operation

DIGITAL INPUT	ANALOG OUTPUT
1 1 1 1 1 1 1 1 1 1	$-V_{REF} (1 - 2^{-10})$
1 0 0 0 0 0 0 0 0 1	$-V_{REF} (1/2 + 2^{-10})$
1 0 0 0 0 0 0 0 0 0	$-V_{REF}/2$
0 1 1 1 1 1 1 1 1 1	$-V_{REF} (1/2 - 2^{-10})$
0 0 0 0 0 0 0 0 0 1	$-V_{REF} (2^{-10})$
0 0 0 0 0 0 0 0 0 0	0

NOTE:

Op-Amp characteristics effect D/A accuracy and settling time. The following Op-Amps, listed in order of increasing speed, are suggested:

1. LM101A   2. LF156A   3. LM118

SILICONIX

Fig. 3-14

FAST VOLTAGE OUTPUT D/A CONVERTER

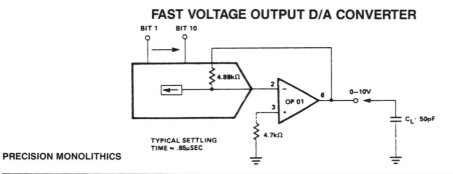

PRECISION MONOLITHICS

Fig. 3-15

10-BIT 4-QUADRANT MULTIPLEXING
D/A CONVERTER (OFFSET BINARY CODING)

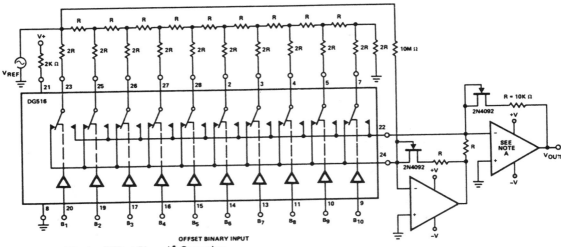

Bipolar (Offset Binary)* Operation

DIGITAL INPUT	ANALOG OUTPUT
1 1 1 1 1 1 1 1 1 1	$-V_{REF} (1 - 2^{-9})$
1 0 0 0 0 0 0 0 0 1	$-V_{REF} (2^{-9})$
1 0 0 0 0 0 0 0 0 0	0
0 1 1 1 1 1 1 1 1 1	$V_{REF} (2^{-9})$
0 0 0 0 0 0 0 0 0 1	$V_{REF} (1 - 2^{-9})$
0 0 0 0 0 0 0 0 0 0	V_{REF}

NOTE: 1 LSB = $2^{-9} V_{REF}$

*Complementing B_1 (MSB) will give 2's complement coding.

SILICONIX

Fig. 3-16

8-BIT D/A CONVERTER II

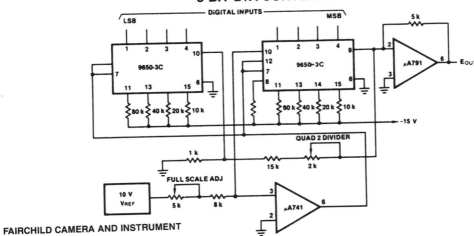

FAIRCHILD CAMERA AND INSTRUMENT

Fig. 3-17

8-BIT D/A WITH OUTPUT CURRENT-TO-VOLTAGE CONVERSION

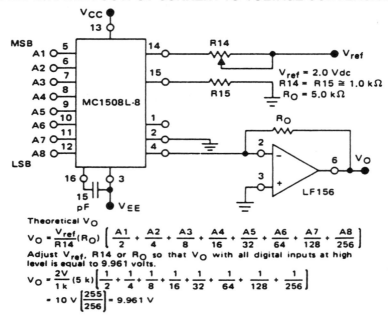

$V_{ref} = 2.0\ Vdc$
$R14 = R15 \cong 1.0\ k\Omega$
$R_O = 5.0\ k\Omega$

Theoretical V_O

$$V_O = \frac{V_{ref}}{R14}(R_O)\left[\frac{A1}{2} + \frac{A2}{4} + \frac{A3}{8} + \frac{A4}{16} + \frac{A5}{32} + \frac{A6}{64} + \frac{A7}{128} + \frac{A8}{256}\right]$$

Adjust V_{ref}, R14 or R_O so that V_O with all digital inputs at high level is equal to 9.961 volts.

$$V_O = \frac{2V}{1\,k}(5\,k)\left[\frac{1}{2} + \frac{1}{4} + \frac{1}{8} + \frac{1}{16} + \frac{1}{32} + \frac{1}{64} + \frac{1}{128} + \frac{1}{256}\right]$$

$$= 10\ V\left[\frac{255}{256}\right] = 9.961\ V$$

MOTOROLA

Fig. 3-18

16-BIT BINARY DAC

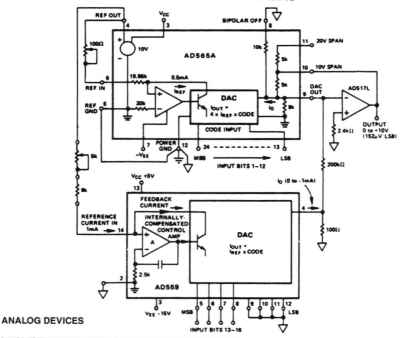

ANALOG DEVICES

Fig. 3-19

DIGITAL-TO-ANALOG CONVERTERS

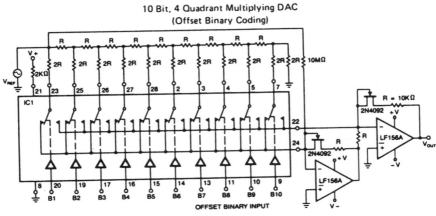

10 Bit, 4 Quadrant Multiplying DAC
(Offset Binary Coding)

IC1: use five of either DG403, DG413, or DG423

Biopolar (Offset Binary)* Operation

DIGITAL INPUT	ANALOG OUTPUT
1 1 1 1 1 1 1 1 1 1	$-V_{REF}\,(1-2^{-9})$
1 0 0 0 0 0 0 0 0 1	$-V_{REF}\,(2^{-9})$
1 0 0 0 0 0 0 0 0 0	0
0 1 1 1 1 1 1 1 1 1	$V_{REF}\,(2^{-9})$
0 0 0 0 0 0 0 0 0 1	$V_{REF}\,(1-2^{-9})$
0 0 0 0 0 0 0 0 0 0	V_{REF}

NOTE: 1 LSB = $2^{-9}\,V_{REF}$

*Complementing B1 (MSB) will give 2's complement coding.

4 Bit Multiplying Current Switch D/A

IC1: use two of either DG403, DG413, or DG423

TYPICAL FEEDTHROUGH ERROR
OF 2mV p-p FOR V_{REF} = 10 V p-p
AND f = 100 MHz

Fig. 3-20

The following applications circuits are intended to illustrate the following points:

- A 2-kΩ resistor should be in series with V+ to limit supply current with negative ringing of the bit inputs
- Temperature compensation for $R_{DS(on)}$ can be provided in the feedback path of the op amp
- Bipolar reference voltages can be used in all configurations

4-CHANNEL D/A OUTPUT AMPLIFIER

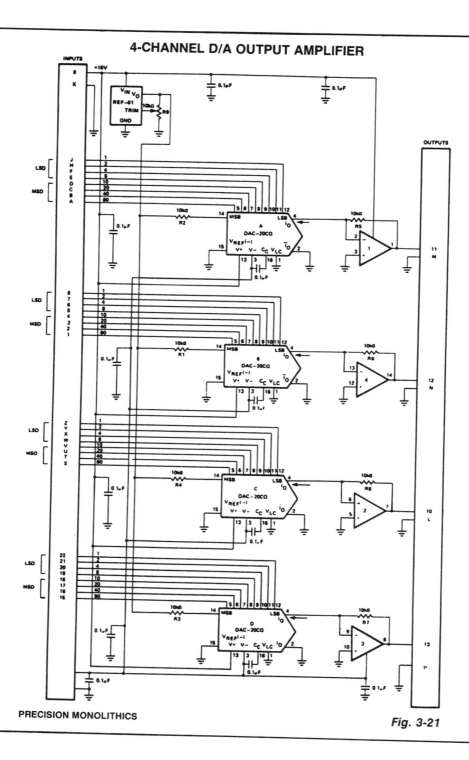

PRECISION MONOLITHICS

Fig. 3-21

4

Filters (Bandpass)

The sources of the following circuits are contained in the Sources section, which begins on page 175. The figure number in the box of each circuit correlates to the source entry in the Sources section.

MULTIPLE-FEEDBACK BANDPASS FILTER

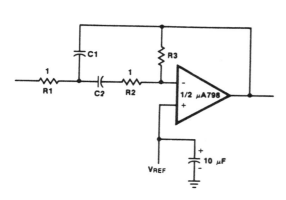

$f_o \overset{\Delta}{=}$ center frequency

$BW \overset{\Delta}{=}$ Bandwidth

R in $k\Omega$

C in μF

$$Q = \frac{f_o}{BW} < 10$$

$$C1 = C2 = \frac{Q}{3}$$

$$\left.\begin{array}{l} R1 = R2 = 1 \\ R3 = 9Q^2 - 1 \end{array}\right\} \text{ Use scaling factors in these expressions}$$

If source impedance is high or varies, filter may be preceded with voltage follower buffer to stabilize filter parameters.

Design example:
given: $Q = 5$, $f_o = 1$ kHz
Let $R1 = R2 = 10$ kΩ
then $R3 = 9(5)^2 - 10$
$R3 = 215$ kΩ
$C = \frac{5}{3} = 1.6$ nF

FAIRCHILD CAMERA AND INSTRUMENT   *Fig. 4-1*

BIQUAD RC ACTIVE BANDPASS FILTER

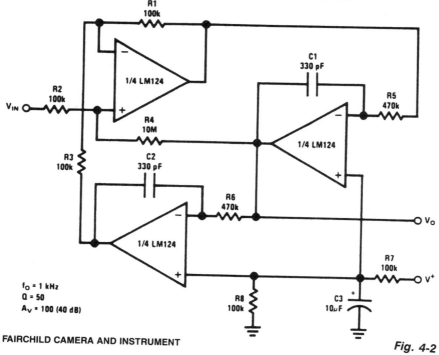

$f_o = 1$ kHz
$Q = 50$
$A_v = 100$ (40 dB)

FAIRCHILD CAMERA AND INSTRUMENT   *Fig. 4-2*

1-kHz BANDPASS ACTIVE FILTER

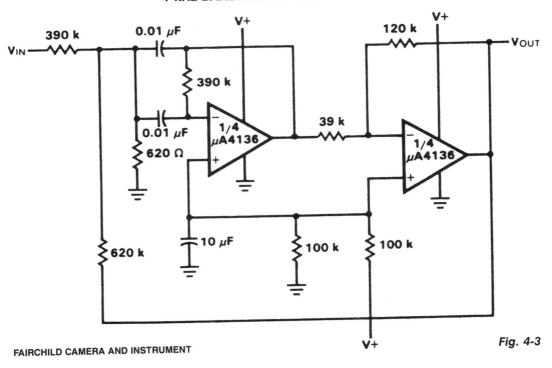

FAIRCHILD CAMERA AND INSTRUMENT

Fig. 4-3

BANDPASS ACTIVE FILTER WITH 60-dB GAIN

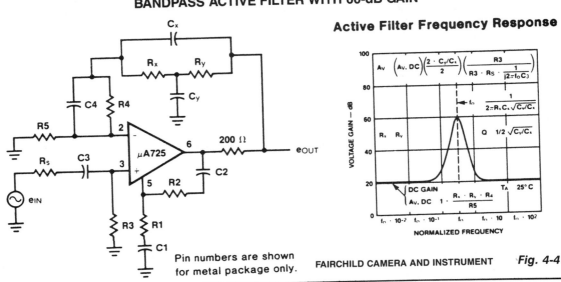

Pin numbers are shown
for metal package only.

FAIRCHILD CAMERA AND INSTRUMENT   *Fig. 4-4*

41

MULTIPLE-FEEDBACK BANDPASS FILTER

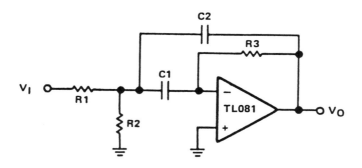

TEXAS INSTRUMENTS

Fig. 4-5

The op amp is connected in the inverting mode. Resistor R3 from the output to the inverting input sets the gain and current through the frequency-determining capacitor, C1. Capacitor C2 provides feedback from the output to the junction of R1 and R2. C1 and C2 are always equal in value. Resistor R2 can be made adjustable in order to adjust the center frequency, which is determined from:

$$f_0 = \frac{1}{2\pi C} \; \frac{1}{R_3} \; \times \; \frac{R_1 + R_2^{1/2}}{R_1 R_2}$$

When designing a filter of this type it is best to select a value for C1 and C2, keeping them equal. Typical audio filters have capacitor values from 0.01 to 0.1 μF, which will result in reasonable values for the resistors.

ACTIVE BANDPASS FILTER

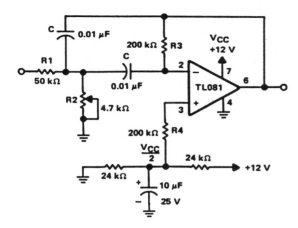

The circuit is a two-pole active filter using a TL081 op amp. This type of circuit is usable only for Qs less than 10. The component values for this filter are calculated from the following equations.

$$R_1 = \frac{Q}{2\,fGC} \qquad\qquad R_3 = \frac{2Q}{2\,fC}$$

$$R_2 = \frac{Q}{(2Q^2 - G)2fC} \qquad\qquad R_4 = R_3$$

The values shown are for a center frequency of 800 Hz.

TEXAS INSTRUMENTS   *Fig. 4-6*

BANDPASS AND NOTCH FILTER

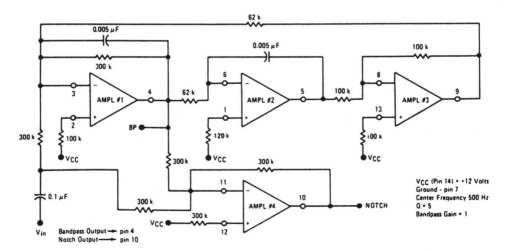

MOTOROLA   *Fig. 4-7*

The Quad op amp MC4301 is used to configure a filter that will notch out a given frequency and produce that notched-out frequency at the BP terminal, useful in communications or measurement setups. By proper component selection, any frequency filter up to a few tens of kilohertz can be obtained.

BANDPASS FILTER

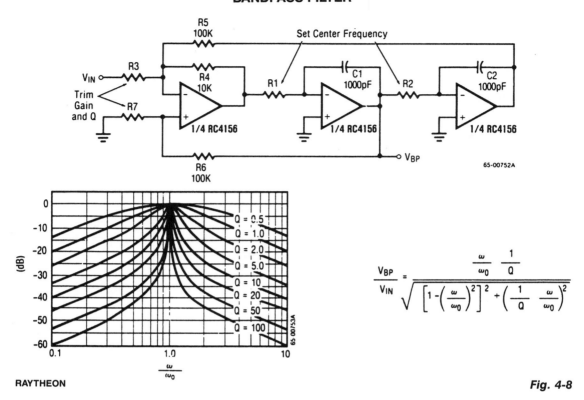

RAYTHEON

Fig. 4-8

The input signal is applied through R3 to the inverting input of the summing amplifier and the output is taken from the first integrator. The summing amplifier will maintain equal voltage at the inverting and non-inverting inputs. Defining $1/R_1C_1$ as ω_1 and $1/R_2C_2$ as ω_2, this is now a convenient form to look at the center-frequency ω_0 and filter Q.

$$\omega_0 = \sqrt{0.1\ \omega_1\ \omega_2}$$

$$\text{and } Q = \left[\frac{1 + \dfrac{10^5}{R_7}}{1.1 + \dfrac{10^4}{R_3}} \right] \omega_0$$

$$= 10^{-9}\sqrt{0.1 R_1 R_2}$$

The frequency response for various values of Q is shown.

44

PROGRAMMABLE BANDPASS USING TWIN-T BRIDGE

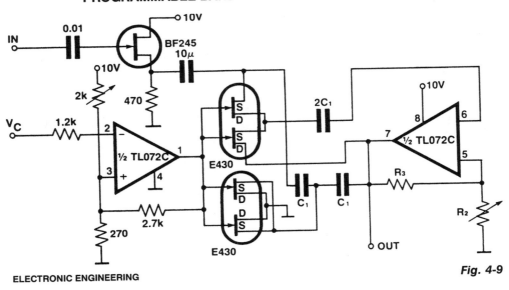

ELECTRONIC ENGINEERING

Fig. 4-9

The circuit gives a programmable bandpass where both the cut-over frequency and the gain, A, are controlled independently. In the twin-T bridge, resistors R and R2 are replaced by two double FETs, E 430, the channel resistance of the first one in the series, the channel resistances of the second one are in parallel as to stimulate the resistance R_2. Both these resistors are controlled by V_c, which ranges from 0 V to about 1 V. The gain of the circuit is set by means of the resistors R2 and R3.

ACTIVE BANDPASS FILTER ($f_0 = 1000$ Hz)

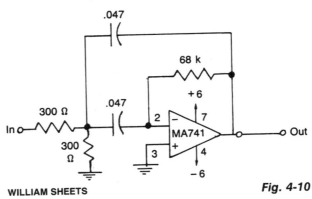

WILLIAM SHEETS

Fig. 4-10

This filter has a bandpass centered around 1 kHz, for applications such as bridge amplifiers, null detectors, etc.

The circuit uses a μA741 IC and standard 5% tolerance components.

HIGH-Q BANDPASS FILTER

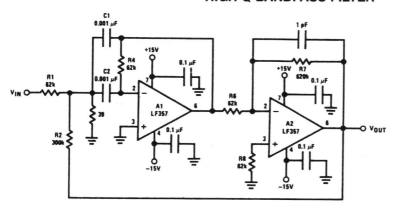

- By adding positive feedback (R2) Q increases to 40
- $f_{BP} = 100\ kHz$

$$\frac{V_{OUT}}{V_{IN}} = 10\sqrt{Q}$$

- Clean layout recommended
- Response to a 1 Vp-p tone burst: 300 μs

Fig. 4-11

SPEECH FILTER (300-Hz BANDPASS)

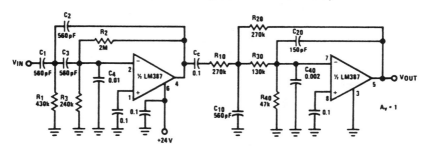

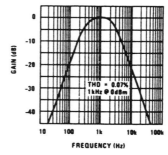

Speech Filter Frequency Response

Fig. 4-12

MULTIPLE-FEEDBACK BANDPASS FILTER (1.0 kHz)

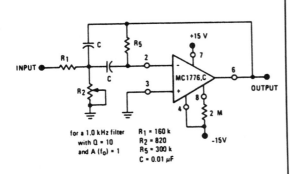

for a 1.0 kHz filter
with Q = 10
and A (f_o) = 1

R_1 = 160 k
R_2 = 820
R_5 = 300 k
C = 0.01 µF

MOTOROLA

Fig. 4-13

TYPICAL ACTIVE BANDPASS FILTER

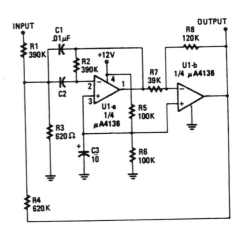

POPULAR ELECTRONICS

Fig. 4-14

20-kHz BANDPASS ACTIVE FILTER

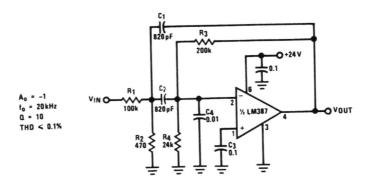

A_o = –1
f_o = 20 kHz
Q = 10
THD < 0.1%

NATIONAL SEMICONDUCTOR

Fig. 4-15

500-Hz SALLEN-KEY BANDPASS FILTER

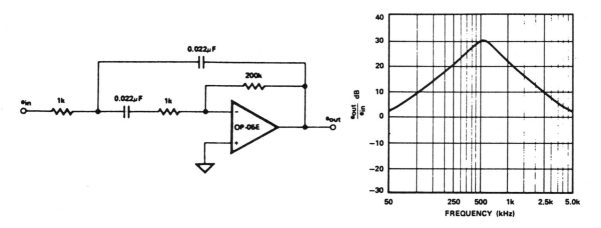

PRECISION MONOLITHICS

Fig. 4-16

SECOND-ORDER BIQUAD BANDPASS FILTER

Note that I_Q on each amplifier might be different. $A_{VCL} = 10$, $Q = 100$, $f_o = 100$ Hz.

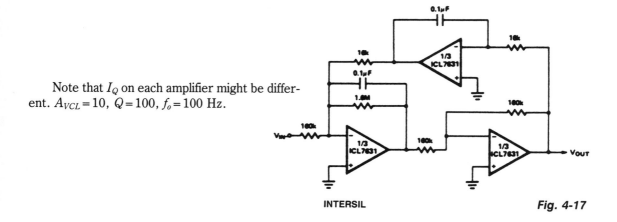

INTERSIL

Fig. 4-17

MFB BANDPASS FILTER FOR MULTICHANNEL TONE DECODER

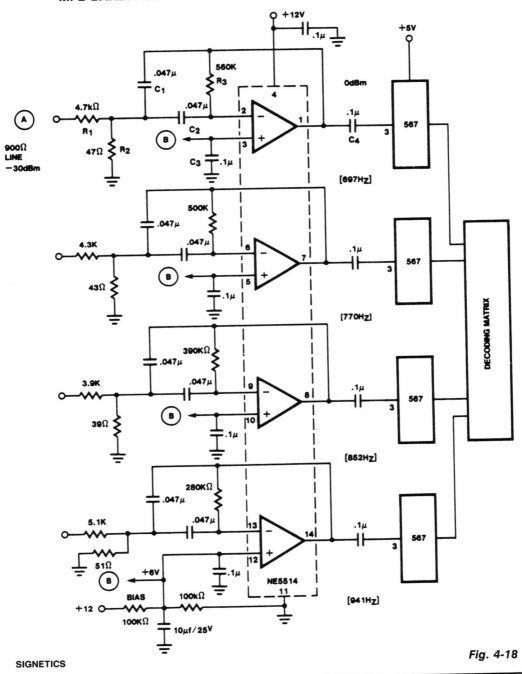

SIGNETICS

Fig. 4-18

49

160-Hz BANDPASS FILTER

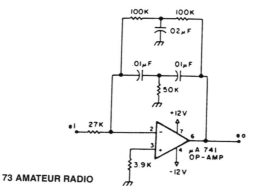

73 AMATEUR RADIO

Fig. 4-19

0.1- TO 10-Hz BANDPASS FILTER

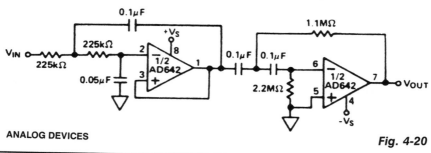

ANALOG DEVICES

Fig. 4-20

VARIABLE-BANDWIDTH ACTIVE BANDPASS FILTER

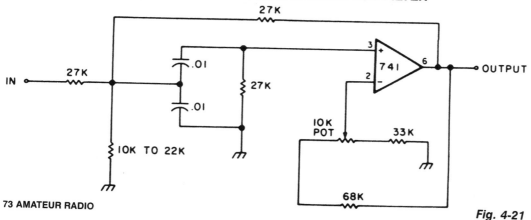

73 AMATEUR RADIO

Fig. 4-21

This circuit has adjustable bandwidth with values for a center frequency of about 800 Hz. The 10-kΩ pot adjusts bandwidth from approximately ±350 to ±140 Hz at 3-dB down points.

FOURTH-ORDER
CHEBYSHEV BANDPASS FILTER

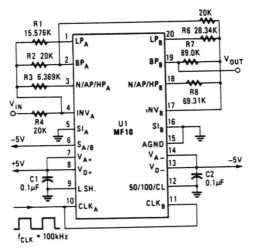

Fig. 4-22

BANDPASS STATE-VARIABLE FILTER

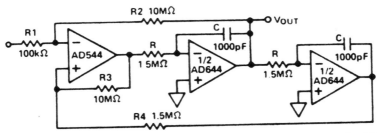

f_o = CENTER FREQUENCY = $1/2 \pi RC$

Q_o = QUALITY FACTOR = $\dfrac{R_1 + R_2}{2R_1}$

H_o = GAIN AT RESONANCE = R_2/R_1

$R_3 = R_4 \approx 10^8/f_o$

Q_o, IS ADJUSTABLE BY VARYING R2

f_o, IS ADJUSTABLE BY VARYING R OR C

Fig. 4-23

5

Filters (High-Pass)

The sources of the following circuits are contained in the Sources section, which begins on page 175. The figure number in the box of each circuit correlates to the source entry in the Sources section.

Fourth-Order High-Pass Butterworth Filter
Wideband Two-Pole High-Pass Filter
High-Pass Active Filter
Second-Order High-Pass Active Filter

High-Frequency High-Pass Filter
Sixth-Order Elliptic High-Pass Filter
Fourth-Order Chebyshev High-Pass Filter

FOURTH-ORDER HIGH-PASS BUTTERWORTH FILTER

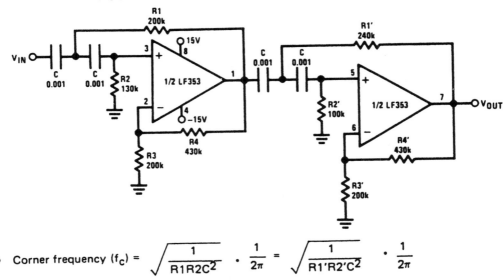

- Corner frequency (f_c) = $\sqrt{\dfrac{1}{R1R2C^2}} \cdot \dfrac{1}{2\pi} = \sqrt{\dfrac{1}{R1'R2'C^2}} \cdot \dfrac{1}{2\pi}$

- Passband gain (H_O) = $(1 + R4/R3)(1 + R4'/R3')$
- First stage Q = 1.31
- Second stage Q = 0.541
- Circuit shown uses closest 5% tolerance resistor values for a filter with a corner frequency of 1 kHz and a passband gain of 10

NATIONAL SEMICONDUCTOR

Fig. 5-1

WIDEBAND TWO-POLE HIGH-PASS FILTER

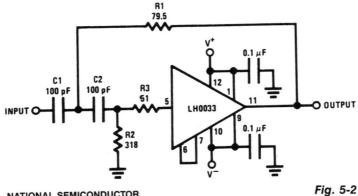

NATIONAL SEMICONDUCTOR

Fig. 5-2

The circuit provides a 10-MHz cutoff frequency. Resistor R3 ensures that the input capacitance of the amplifier does not interact with the filter response at the frequency of interest. An equivalent low-pass filter is similarly obtained by capacitance and resistance transformation.

HIGH-PASS ACTIVE FILTER

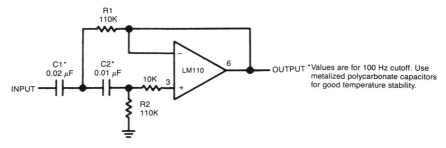

*Values are for 100 Hz cutoff. Use metalized polycarbonate capacitors for good temperature stability.

NATIONAL SEMICONDUCTOR

Fig. 5-3

SECOND-ORDER
HIGH-PASS ACTIVE FILTER

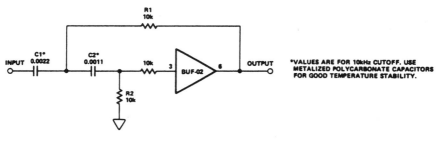

*VALUES ARE FOR 10kHz CUTOFF. USE METALIZED POLYCARBONATE CAPACITORS FOR GOOD TEMPERATURE STABILITY.

PRECISION MONOLITHICS

Fig.5-4

HIGH-FREQUENCY
HIGH-PASS FILTER

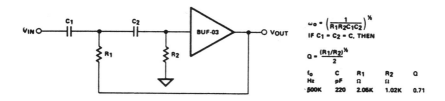

$$\omega_o = \left(\frac{1}{R_1 R_2 C_1 C_2} \right)^{\frac{1}{2}}$$

IF $C_1 = C_2 = C$, THEN

$$Q = \frac{(R_1/R_2)^{\frac{1}{2}}}{2}$$

f_o Hz	C pF	R_1 Ω	R_2 Ω	Q
500K	220	2.05K	1.02K	0.71

PRECISION MONOLITHICS

Fig. 5-5

SIXTH-ORDER ELLIPTIC HIGH-PASS FILTER

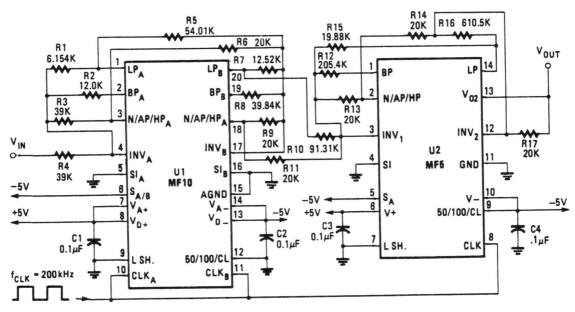

Fig. 5-6

FOURTH-ORDER
CHEBYSHEV HIGH-PASS FILTER

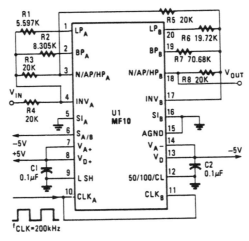

Fig. 5-7

6

Filters (Low-Pass)

The sources of the following circuits are contained in the Sources section, which begins on page 175. The figure number in the box of each circuit correlates to the source entry in the Sources section.

PRECISION FAST-SETTLING LOW-PASS FILTER

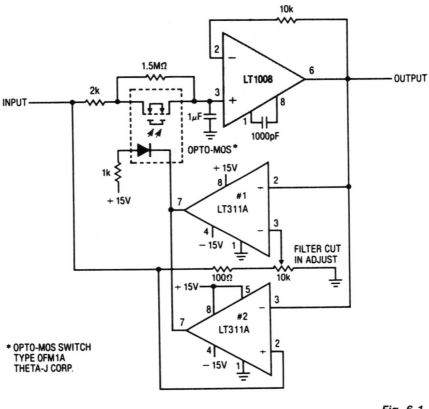

LINEAR TECHNOLOGY

Fig. 6-1

This circuit is useful where fast signal acquisition and high precision are required, as in electronic scales. The filter's time constant is set by the 2-kΩ resistor and 1-μF capacitor, until comparator No. 1 switches. The time constant is then set by the 1.5-MΩ resistor and the 1-μF capacitor. Comparator No. 2 provides a quick reset. The circuit settles to a final value three times as fast as a simple 1.5-MΩ /1-μF filter, with almost no dc error.

ACTIVE LOW-PASS FILTER WITH DIGITALLY SELECTED BREAK FREQUENCY

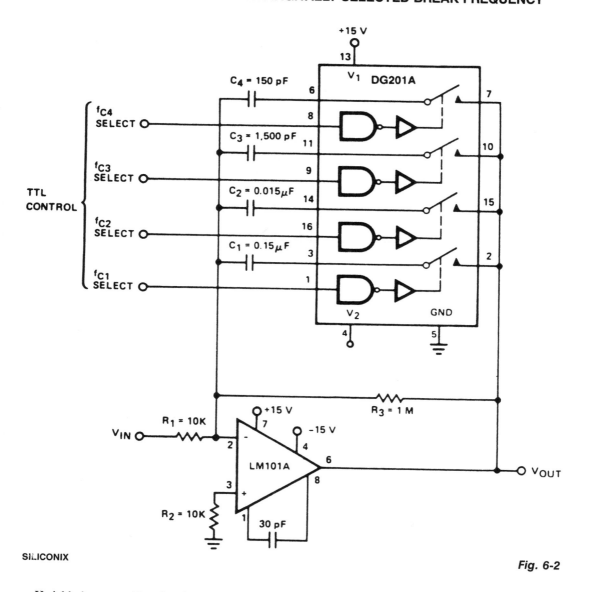

SILICONIX

Fig. 6-2

Variable low-pass filter has break frequencies at 1, 10, 100 Hz, and 1 kHz. The break frequency is:

$$f_c = \frac{1}{2\,\pi\,R_3\,C_X}$$

ACTIVE LOW-PASS FILTER WITH DIGITALLY SELECTED BREAK FREQUENCY *(Cont.)*

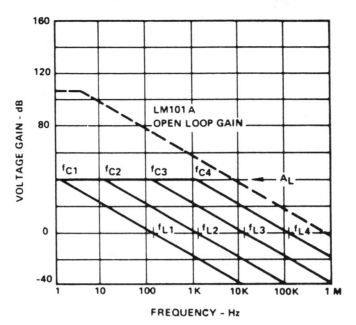

A_L (VOLTAGE GAIN BELOW BREAK FREQUENCY)

$$= \frac{R_3}{R_1} = 100 \ (40 \ dB)$$

f_c (BREAK FREQUENCY) $= \dfrac{1}{2\pi R_3 C_X}$

f_L (UNITY GAIN FREQUENCY) $= \dfrac{1}{2\pi R_1 C_X}$

MAX ATTENUATION $= \dfrac{r_{DS(on)}}{10K} \approx -40 \ dB$

The low frequency gain is:

$$A_L = \frac{R_3}{R_3} = 100 \ (40 \ \text{dB})$$

A second break frequency (a zero) is introduced by $r_{DS(on)}$ of the DG201A, causing the minimum gain to be:

$$A_{MIN} = \frac{r_{DS(on)}}{R_1} \approx \frac{100}{10 \ \text{k}\Omega} = 0.01$$

a maximum attenuation of 40 dB (80 dB relative to the low-frequency gain).

POLE-ACTIVE LOW-PASS FILTER
(BUTTERWORTH MAXIMALLY FLAT RESPONSE)

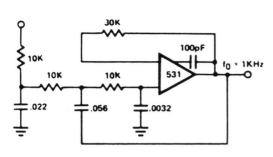

*Reference—EDN Dec. 15, 1970
Simplify 3-Pole Active Filter Design
A. Paul Brokow

SIGNETICS

Fig. 6-3

RESPONSE OF 3-POLE ACTIVE BUTTERWORTH MAXIMALLY FLAT FILTER

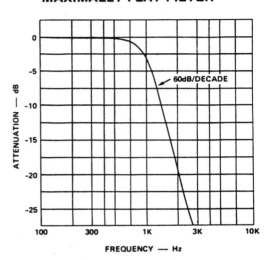

EQUAL-COMPONENT SALLEN-KEY LOW-PASS FILTER

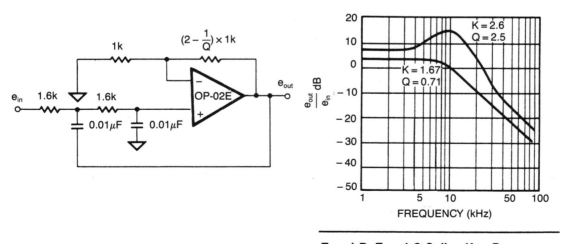

Equal R, Equal C Sallen-Key Response

PRECISION MONOLITHICS

Fig. 6-4

400-Hz LOW-PASS BUTTERWORTH ACTIVE FILTER

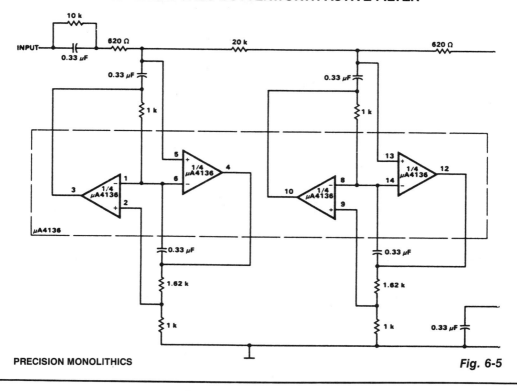

PRECISION MONOLITHICS

Fig. 6-5

10-kHz SALLEN-KEY LOW-PASS FILTER

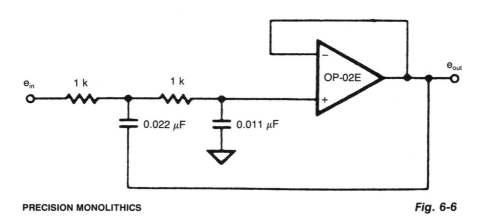

PRECISION MONOLITHICS

Fig. 6-6

SALLEN-KEY SECOND-ORDER LOW-PASS FILTER

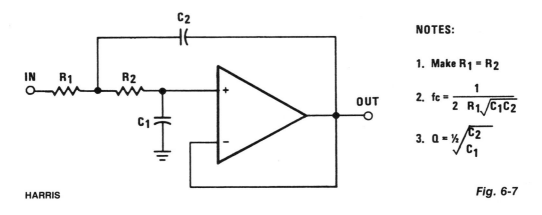

NOTES:

1. Make $R_1 = R_2$

2. $fc = \dfrac{1}{2\,R_1\sqrt{C_1 C_2}}$

3. $Q = \frac{1}{2}\sqrt{\dfrac{C_2}{C_1}}$

HARRIS

Fig. 6-7

FIFTH-ORDER CHEBYSHEV MULTIPLE-FEEDBACK LOW-PASS FILTER

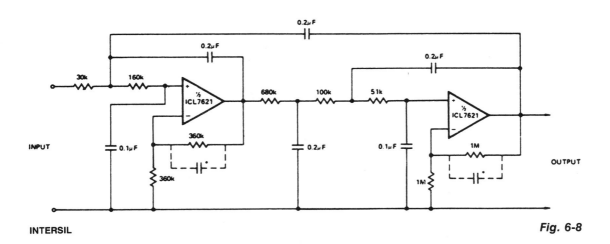

INTERSIL

Fig. 6-8

The low bias currents permit high-resistance and low-capacitance values to be used to achieve low frequency cutoff. $f_c = 10$ Hz, $A_{\text{VCL}} = 4$, Passband ripple = 0.1 dB. Notice that small capacitors (25 to 50 pF) might be needed for stability in some cases.

LOW-PASS FILTER

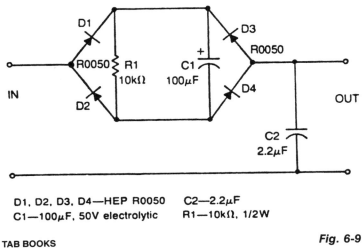

D1, D2, D3, D4—HEP R0050   C2—2.2µF
C1—100µF, 50V electrolytic   R1—10kΩ, 1/2W

TAB BOOKS   *Fig. 6-9*

This nonlinear, passive filter circuit rejects ripple (or unwanted, but fairly steady voltage) without appreciably affecting the rise time of a signal. The circuit works best when the signal level is considerably lower than the unwanted ripple, provided that the ripple level is fairly constant. The circuit has characteristics similar to two peak-detecting sample-and-hold circuits in tandem with a voltage averager.

7

Filters—Miscellaneous

The sources of the following circuits are contained in the Sources section, which begins on page 175. The figure number in the box of each circuit correlates to the source entry in the Sources section.

Programmable Active Filters
Biquad Audio Filter
Low-Cost Universal Active Filter
Biquad Filter
Razor-Sharp CW Filter
Low-Power Active Filter with Digitally

Selectable Center Frequency
Glitch-Free Turbo Circuit
Voltage-Controlled Filter
Digitally Tuned Low-Power Active Filter
Filter Networks

PROGRAMMABLE ACTIVE FILTERS

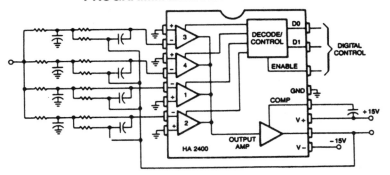

HARRIS

Fig. 7-1

This is a second-order, low-pass filter with programmable cutoff frequency. This circuit should be driven from a low-source impedance because there are paths from the output to the input through the unselected networks. Virtually any filter function that can be constructed with a conventional op amp can be made programmable with the HA-2400.

A useful variation would be to wire one channel as a unity-gain amplifier so that one could select the unfiltered signal, or the same signal filtered in various manners. These could be cascaded to provide a wide variety of programmable filter functions.

BIQUAD AUDIO FILTER

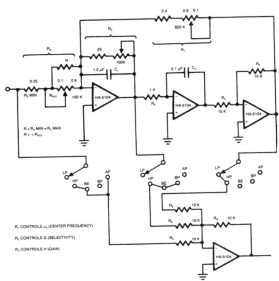

The biquad offers a universal filter with ω_0. Q. and gain "orthogonally" tuned.

HARRIS

Fig. 7-2

This universal filter offers low-pass, high-pass, bandpass, band-elimination, and all-pass functions. The Biquad consists of two successive integration stages followed by an inverting stage. The entire group has a feedback loop from the front to the back consisting of R1, which is chiefly responsible for controlling the center frequency, ω_0. The first stage of integration is a *poor* integrator because R2 limits the range of integration. R2 and C form the time constant of the first stage integrator with R3 influencing gain H almost directly. The bandpass function is taken after the first stage with the low-pass function taken after the third stage. The remaining filter operations are generated by various combinations of three stages.

The Biquad is orthogonally tuned, meaning that ω_0, Q, and gain H can all be independently adjusted. The component values known will allow ω_0 to range from 40 Hz to 20 kHz. The other component values give an adequate range of operation to allow for virtually universal filtering in the audio region. ω_0, Q, and gain H can all be independently adjusted by tuning R1 through R3 in succession.

65

LOW-COST UNIVERSAL ACTIVE FILTER

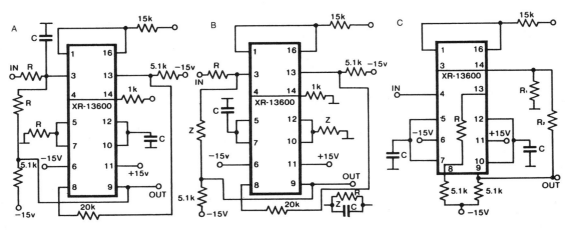

Fig. 7-3

The circuit shown in A gives the bandpass operation the transfer function calculated from:

$$F_{BP}(s) = \frac{S/\omega_0}{K}$$

$$\text{where } K = 1 + \frac{s}{Q\omega_o} + \frac{s^2}{\omega_o^2}$$

The cut-off frequency, ω_o, and the Q-factor are given by

$$\omega_o = \frac{g}{C} \text{ and } Q = \frac{gR}{2}$$

where g is the transconductance at room temperature.

Interchanging the capacitor C with the resistor R at the input of the circuit, high-pass operation is obtained. A low-pass filter is obtained by applying two parallel connections of R and C, as shown in B.

The low-pass operation may be much improved with the circuit in C. Here the gain and Q may be set up separately with respect to the cut-off frequency according to the equations

$$Q = \frac{1}{fB} = 1 + \frac{R_2}{R_1}$$

$$A = Q^2 \text{ and } \omega_o = \frac{g\,fB}{C}$$

BIQUAD FILTER

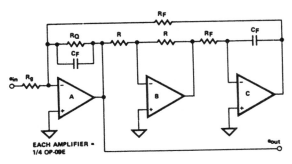

The biquad filter, which appears to be very similar to the state-variable filter, has a bandwidth that is fixed, regardless of the center frequency. This type of filter is useful in applications, such as spectrum analyzers, which require a filter with a fixed bandwidth.

PRECISION MONOLITHICS

Fig. 7-4

RAZOR-SHARP CW FILTER

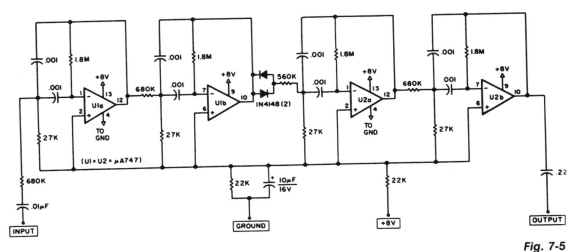

AMATEUR RADIO

Fig. 7-5

The circuit consists of four stages of active bandpass filtering provided by two type-μA747 integrated-circuit dual op amps and includes a simple threshold detector (diodes D1 and D2) between stages 2 and 3 to reduce low-level background noise. Each of the four filter stages acts as a narrow bandpass filter with an audio bandpass centered at 750 Hz. The actual measured 3-dB bandwidth is only 80 Hz wide.

LOW-POWER ACTIVE FILTER WITH DIGITALLY SELECTABLE CENTER FREQUENCY

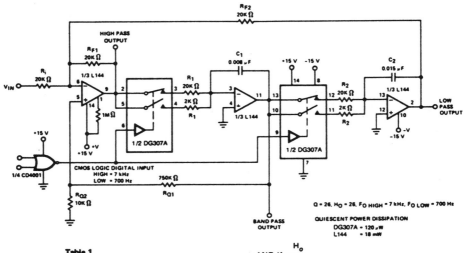

Table 1
Design Procedure for the State Variable Active Filter
Given: f_0 (Resonant Frequency),
H_0 (Gain at the Resonant Frequency) and Q_0

STANDARD DESIGN
(Assumes Infinte Op-Amp Gain)

1. CHOOSE $C_1 = C_2 = C$, A CONVENIENT VALUE

2. LET $R_1 = R_2 = R$

3. THEN $R = \dfrac{1}{2\pi \times f_0 \times C}$

4. CHOOSE $R_{11} = R_{12} = KR$,
   WHERE R_{11}, R_{12} = A CONVENIENT VALUE

AND $K = \dfrac{H_o}{Q_o}$

IF H_o IS UNIMPORTANT (i.e., GAIN CAN BE
ADDED BEFORE AND/OR AFTER THE
FILTER), CHOOSE K = 1

5. LET R_{Q1} = A CONVENIENT VALUE

6. THEN $R_{Q2} = \dfrac{R_{Q1}}{(2 + K) \times Q_o - 1}$

A (f_o) = THE NOMINAL OP AMP GAIN AT
THE RESONANT FREQUENCY.

GBWP = THE NOMINAL GAIN-BANDWIDTH
PRODUCT OF THE OPERATIONAL AMPLIFIER

SILICONIX

Fig. 7-6

The switchable center frequency active filter allows a decade change in center frequency.

GLITCH-FREE TURBO CIRCUIT

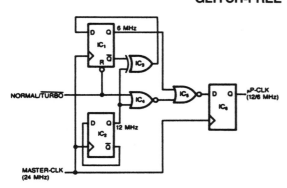

This simple circuit generates a dual-speed clock for personal computers. The circuit synchronizes your asynchronous switch inputs with the master clock to provide glitch-free transitions from one clock speed to the other. The dual-speed clock allows some programs to run at the higher clock speed in order to execute more quickly. Other programs—for example, programs that use loops for timing—can still run at the lower speed as necessary. The circuit will work with any

EDN

Fig. 7-7

master-clock frequency that meets the flip-flops' minimum-pulse-width specs.

The two D flip-flops, IC1 and IC2, and an XOR gate, IC3, form a binary divider that develops the 6- and 12-MHz clocks. When the NT signal is low, the reset pin forces the 6-MHz output low. On the other hand, when the NT signal is high, IC3 blocks the 12-MHz output. Therefore, only one of the two clock signals passes through IC3 and gets clocked into IC6. Because the master-clk signal clocks IC6, asynchronous switching of the NT signal can't generate an output pulse shorter than 41 μs ($^1/_{24}$ MHz). Also, the synchronization eliminates glitches.

VOLTAGE-CONTROLLED FILTER

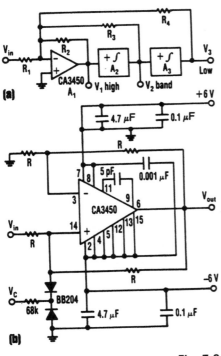

(a)

(b)

ELECTRONIC DESIGN   *Fig. 7-8*

The control voltage V_C easily sets the cut-off frequency ω_o of this state-variable filter to any desired value, from about 1.7 MHz up to 5 MHz, with a BB 204 varicap and $R = 100$ kΩ. V_C can range from 0 to 28 V. This range changes the capacitance of the varicap from about 4 to 12 pF.

The circuit consists of input summing circuit A1 and two noninverting integrators, A2 and A3. Both the integrators and the summing-amplifier circuits use CA3450 op amps. With them, cut-off frequencies up to 200 MHz are possible.

The circuit's cut-off frequency, its Q-factor, and gain G are simply:

$$\omega_o = 2/CR, \quad Q = R_3/R_4,$$
$$\text{and } G = R_4/R_1$$

For a given value for R_4, say 10 kΩ, Q depends only upon the resistance of R_3. The Q can be any value, even 100, independently of both ω_o and G. Similarly, the gain then depends only on the resistance of R_1. It can be set as high as 100.

DIGITALLY TUNED LOW-POWER ACTIVE FILTER

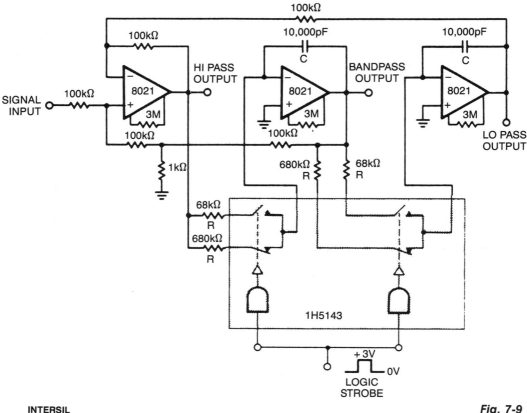

INTERSIL

Fig. 7-9

Constant gain, constant Q, variable frequency filter which provides simultaneous low-pass, band-pass, and high-pass outputs. With the component values shown, center frequency will be 235 Hz and 23.5 Hz for high and low logic inputs, respectively, $Q=100$, and gain $=100$.

$$f_n = \text{center frequency} = \frac{1}{2\pi RC}$$

FILTER NETWORKS

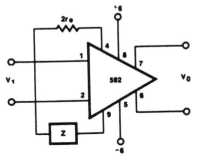

$$\frac{V_0\,(s)}{V_1\,(s)} \cong \frac{1.4 \times 10^4}{Z(s) + 2r_e}$$

$$\cong \frac{1.4 \times 10^4}{Z(s) + 32}$$

BASIC CONFIGURATION

Z NETWORK	FILTER TYPE	$\dfrac{V_0\,(s)}{V_1\,(s)}$ TRANSFER FUNCTION
—R—L—	LOW PASS	$\dfrac{1.4 \times 10^4}{L}\left[\dfrac{1}{s + R/L}\right]$
—R—C—	HIGH PASS	$\dfrac{1.4 \times 10^4}{R}\left[\dfrac{s}{s + 1/RC}\right]$
—R—L—C—	BAND PASS	$\dfrac{1.4 \times 10^4}{L}\left[\dfrac{s}{s^2 + R/L\,s + 1/LC}\right]$
—R— (L ∥ C)	BAND REJECT	$\dfrac{1.4 \times 10^4}{R}\left[\dfrac{s^2 + 1/LC}{s^2 + 1/LC + s/RC}\right]$

NOTE
In the networks above, the R value used is assumed to include $2r_e$, or approximately 32Ω.

Fig. 7-10

8

Filters (Noise)

The sources of the following circuits are contained in the Sources section, which begins on page 175. The figure number in the box of each circuit correlates to the source entry in the Sources section.

Rumble Filter   Dynamic Noise Filter
Scratch Filter I   Noisy Signals Filter
Scratch Filter II   Tunable Notch Filter

RUMBLE FILTER

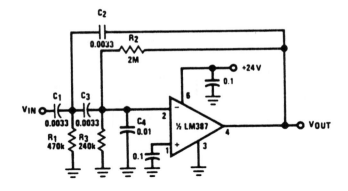

$f_c = 50\,Hz$

SLOPE = −12 dB/OCTAVE

$A_0 = -1$

THD < 0.1%

NATIONAL SEMICONDUCTOR

Fig. 8-1

SCRATCH FILTER I

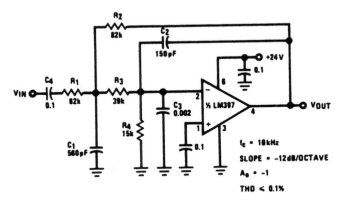

$f_c = 10\,kHz$

SLOPE = −12 dB/OCTAVE

$A_0 = -1$

THD < 0.1%

NATIONAL SEMICONDUCTOR

Fig. 8-2

SCRATCH FILTER II

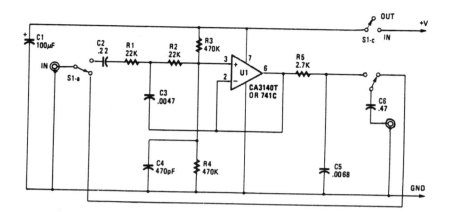

POPULAR ELECTRONICS

Fig. 8-3

DYNAMIC NOISE FILTER

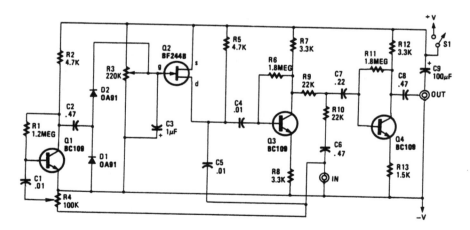

POPULAR ELECTRONICS

Fig. 8-4

NOISY SIGNALS FILTER

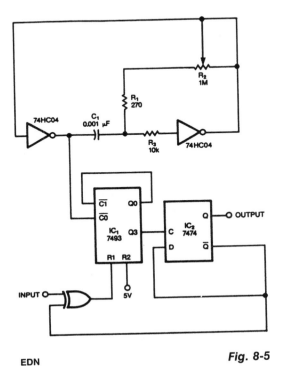

This circuit filters noise, such as glitches and contact bounce, from digital signals. You can easily adjust the circuit for a wide range of noise frequencies. The circuit's output changes state only if the input differs from the output long enough for the counter to count eight cycles. If the input changes before the counter reaches its maximum count, the counter resets without clocking the output of flip-flop, IC2. Use R2 to set the frequency of the two-inverter CMOS oscillator, which clocks the counter. Simply adjust the oscillator so that its period is one-eighth that of the noise you want to eliminate.

Fig. 8-5

TUNABLE NOTCH FILTER

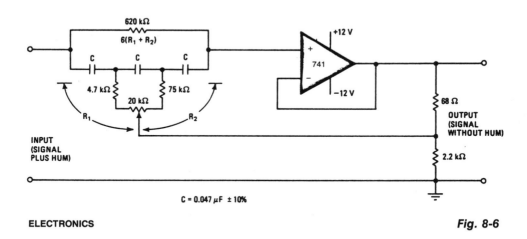

ELECTRONICS

Fig. 8-6

This narrow-stop-band filter can be tuned by the pot to place the notch at any frequency from 45 to 90 Hz. It attenuates power-line hum or other unwanted signals by at least 30 dB. Because the circuit uses wide-tolerance parts, it is inexpensive to build.

9

Filters (Notch)

The sources of the following circuits are contained in the Sources section, which begins on page 175. The figure number in the box of each circuit correlates to the source entry in the Sources section.

Adjustable Q Notch Filter
1800-Hz Notch Filter
550-Hz Notch Filter
Tunable Audio Notch Filter
Audio Notch Filter
Tunable Notch Filter
Active Band-Reject Filter
Wien-Bridge Notch Filter
Tunable Audio Filter
Passive-Bridged Differentiator-Tunable
    Notch Filter

Three-Amplifier Notch Filter (or Elliptic
    Filter Building Block)
Selectable-Bandwidth Notch Filter
High-Q Notch Filter
Rejection Filter
Notch Filter
Twin-T Notch Filter
4.5-MHz Notch Filter

ADJUSTABLE Q NOTCH FILTER

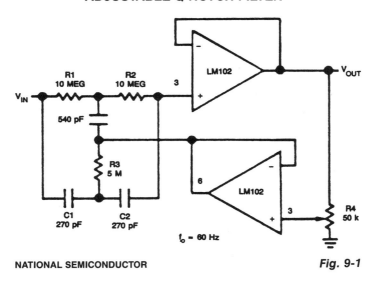

NATIONAL SEMICONDUCTOR

Fig. 9-1

In applications where the rejected signal might deviate slightly from the null on the notch network, it is advantageous to lower the Q of the network. This ensures some rejection over a wider range of input frequencies. The figure shows a circuit where the Q can be varied from 0.3 to 50. A fraction of the output is fed back to R3 and C3 by a second voltage follower, and the notch, Q, is dependent on the amount of signal fed back. A second follower is necessary to drive the twin "T" from a low-resistance source so that the notch frequency and depth will not change with the potentiometer setting.

1800-Hz NOTCH FILTER

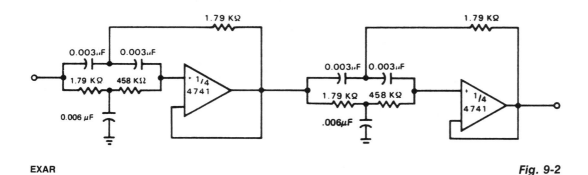

EXAR

Fig. 9-2

The circuit produces at least 60 dB of attenuation at the notch frequency.

550-Hz NOTCH FILTER

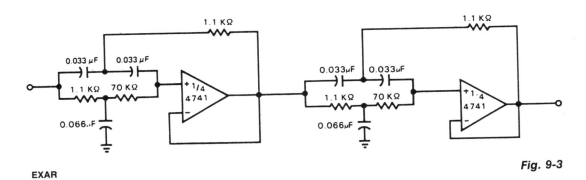

EXAR

Fig. 9-3

The circuit produces at least 60 dB of attenuation at the notch frequency.

TUNABLE AUDIO NOTCH FILTER

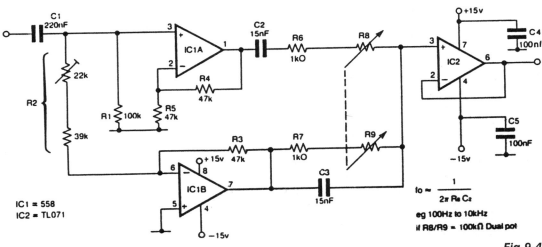

IC1 = 558
IC2 = TL071

$$fo \approx \frac{1}{2\pi R_8 C_2}$$

eg 100Hz to 10kHz
if R8/R9 = 100kΩ Dual pot

ELECTRONIC ENGINEERING

Fig.9-4

The circuit requires only one dual-ganged potentiometer to tune over a wide range; if necessary, it can tune over the entire audio range in one sweep. The principle used is that of the Wien bridge, fed from anti-phase inputs. The output should be buffered (as shown) with a FET input op amp, particularly if a high-value pot is used. An op amp with differential outputs (e.g., MC1445) can be used in place of the driver ICs; R2 can be made trimmable to optimize the notch.

AUDIO NOTCH FILTER

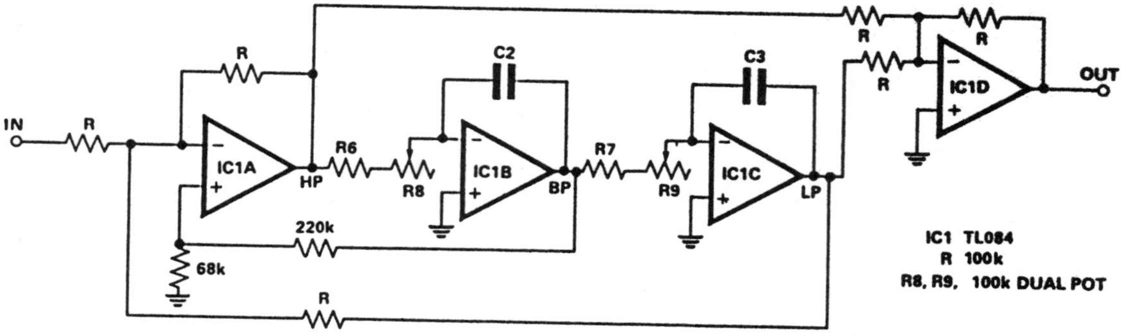

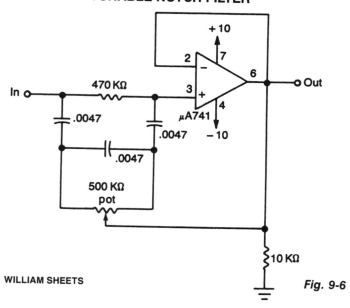

ELECTRONIC ENGINEERING

Fig. 9-5

With the circuit shown here, the response at one octave off tune is within 10% of the far out response: notch sharpness can be increased or reduced by reducing or increasing respectively the 68-kΩ resistor. Linearity tracking of R8 and R9 has no effect on the notch depth. The signals at HP and LP are always in antiphase; notch will always be very deep at the tuned frequency, despite tolerance variations in R6, R7, R8, R9, C2, and C3.

TUNABLE NOTCH FILTER

WILLIAM SHEETS

Fig. 9-6

This notch filter is useful for tunable band-reject applications in the audio range. The values shown will give a tuning range of about 300 to 1500 Hz.

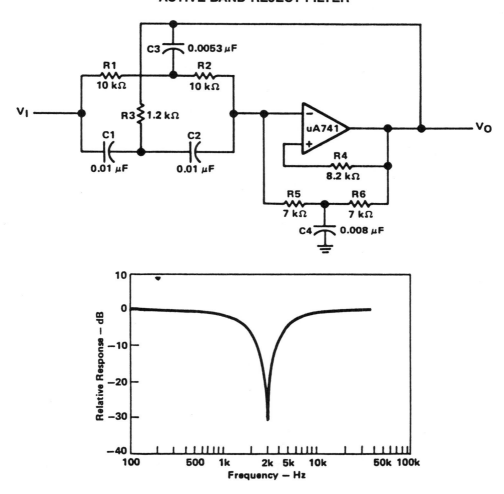

ACTIVE BAND-REJECT FILTER

TEXAS INSTRUMENTS

Fig. 9-7

A filter with a band-reject characteristic is frequently referred to as a *notch filter*. A typical circuit using a μA 741 is the unity-gain configuration for this type of active filter shown. The filter response curve shown is a second-order band-reject filter with a notch frequency of 3 kHz. The resulting Q of this filter is about 23, with a notch depth of -31 dB. Although three passive T networks are used in this application, the operational amplifier has become a sharply tuned low-frequency filter without the use of inductors or large-value capacitors.

WIEN-BRIDGE NOTCH FILTER

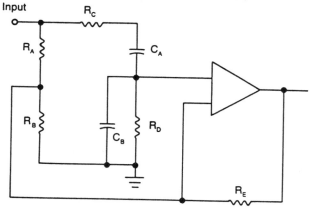

Input

WILLIAM SHEETS

Fig. 9-8

If $R_A = R_B = R_C = R_D = R_E = R$ and $C_A = C_B = C$

$$f_{null} = \frac{1}{6.28\ RC}\quad \begin{matrix} R\ M\Omega \\ C\ \mu F \\ f\ Hz \end{matrix}$$

TUNABLE AUDIO FILTER

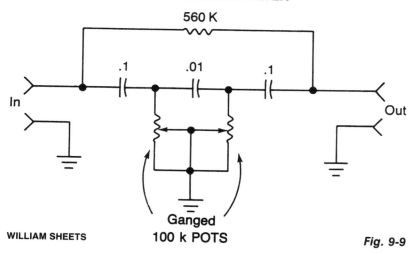

WILLIAM SHEETS

Ganged
100 k POTS

Fig. 9-9

This filter covers the upper part of the audio passband and can be used to eliminate unwanted high frequencies from audio signals.

PASSIVE-BRIDGED DIFFERENTIATOR-TUNABLE NOTCH FILTER

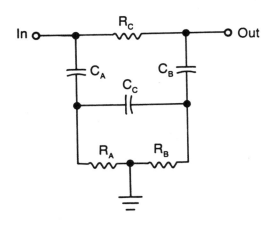

If $C_A = C_B = C_C = C$
And if $R_3 = 6(R_A + R_B)$

$$\text{Then } \left\{ \begin{array}{c} \text{Notch} \\ \text{freq} \end{array} \right\} = \frac{1}{6.28 \ C\sqrt{3R_A R_B}}$$

If R_A and R_B are made a potentiometer, then the filter can be variable.

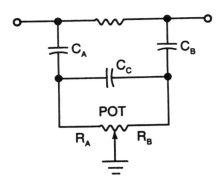

WILLIAM SHEETS   *Fig. 9-10*

R_A and R_B are sections of potentiometer.

THREE-AMPLIFIER NOTCH FILTER (OR ELLIPTIC FILTER BUILDING BLOCK)

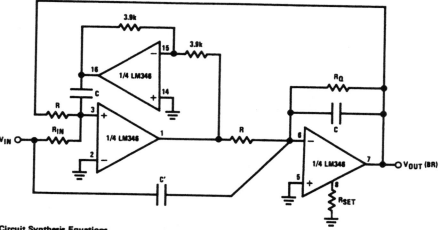

Circuit Synthesis Equations

$$R \times C = \frac{0.159}{f_o} \; ; R_Q = Q_o \times R; R_{IN} = \frac{0.159 \times f_o}{C' \times f^2_{notch}}$$

• For nothing but a notch output: $R_{IN} = R$, $C' = C$.

$$H_{o(BR)} \Big|_{f \ll f_{notch}} = \frac{R}{R_{IN}} \qquad H_{o(BR)} \Big|_{f \gg f_{notch}} = \frac{C'}{C}$$

NATIONAL SEMICONDUCTOR

Fig. 9-11

SELECTABLE-BANDWIDTH NOTCH FILTER

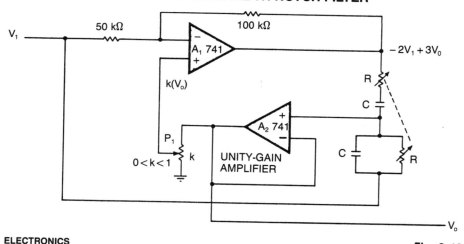

ELECTRONICS

Fig. 9-12

This notch filter, which operates at up to 200 kHz, uses a modified Wien bridge to select bandwidth over which frequencies are rejected. RC components determine filter's center frequency, P1 selects notch bandwidth. Notch depth is fixed at about 60 dB.

HIGH-Q NOTCH FILTER

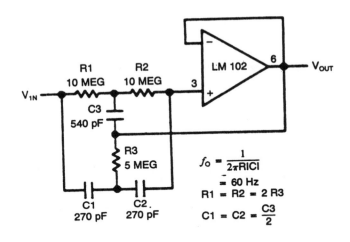

$$f_o = \frac{1}{2\pi RICI}$$
$$\cong 60 \text{ Hz}$$
$$R1 = R2 = 2\,R3$$
$$C1 = C2 = \frac{C3}{2}$$

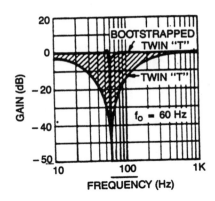

Response of High and Low Q Notch Filter

NATIONAL SEMICONDUCTOR   *Fig. 9-13*

   This circuit shows a twin-T network connected to an LM102 to form a high-Q, 60-Hz notch filter. The junction of R3 and C3 which is normally connected to ground, is bootstrapped to the output of the follower. Because the output of the follower is a very low impedance, neither the depth nor the frequency of the notch change; however, the Q is raised in proportion to the amount of signal fed back to R3 and C3. Shown is the response of a normal twin-T and the response with the follower added.

REJECTION FILTER

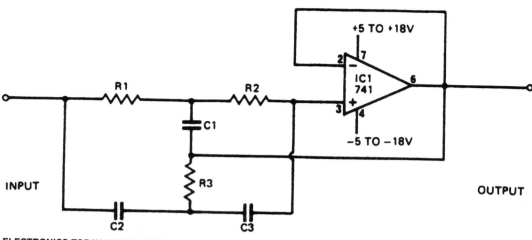

ELECTRONICS TODAY INTERNATIONAL

Fig. 9-14

This narrowband filter using the 741 operational amplifier can provide up to 60 dB of rejection. With resistors equal to 100 kΩ and capacitors equal to 320 pF, the circuit will reject 50 Hz. Frequencies within the range 1 Hz to 10 kHz can be rejected by selecting components in accordance with the formula:

$$F = \frac{1}{2\pi RC}$$

To obtain rejections better than 40 dB, resistors should be matched to 0.1% and capacitors to 1%.

NOTCH FILTER

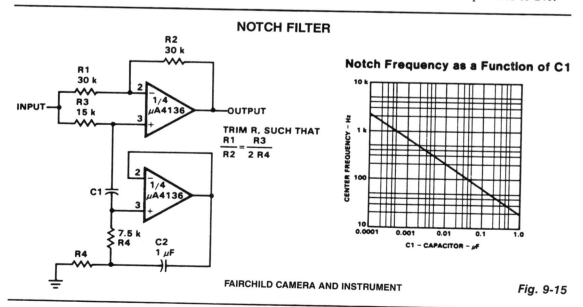

FAIRCHILD CAMERA AND INSTRUMENT

Fig. 9-15

TWIN-T NOTCH FILTER

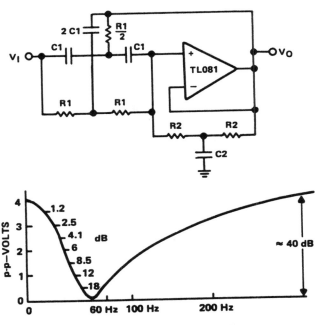

60-Hz Twin-T Notch Filter Response   *Fig. 9-16*

This filter is used to reject or block a frequency or band of frequencies. These filters are often designed into audio and instrumentation systems to eliminate a single frequency, such as 60 Hz. Commercial grade components with 5 to 10% tolerance produce a null depth of at least 30 to 40 dB. When a twin-T network is combined with a TL081 op amp in a circuit, an active filter can be implemented. The added resistor capacitor network, R2 and C2, work effectively in parallel with the original twin-T network, on the input of the filter. These networks set the Q of the filter. The op amp is basically connected as a unity-gain voltage follower. The Q is found from:

$$Q = \frac{R_2}{2R_1} = \frac{C_1}{C_2}$$

For a 60-Hz notch filter with a Q of 5, it is usually best to pick the C_1 capacitance and calculate resistance R_1. Let $C_1 = 0.22\ \mu F$. Then:

$$R_1 = 12\ k\Omega$$
$$R_2 = 120\ k\Omega$$
$$C_2 = 0.047 \mu F$$

Standard 5% resistors and 10% capacitors produce a notch depth of about 40 dB, as shown in the frequency-response curve.

4.5-MHz NOTCH FILTER

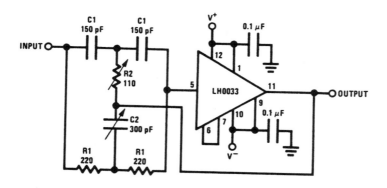

NATIONAL SEMICONDUCTOR

Fig. 9-17

Component-value sensitivity is extremely critical, as are temperature coefficients and matching of the components. Best performance is attained when perfectly matched components are used and when the gain of the amplifier is unity. To illustrate, the quality factor Q is very high as amplifier gain approaches 1 with all components matched (in fact, theoretically it approaches ∞) but decreases to about 12.5 with the amplifier gain at 0.98.

10

Filters (State Variable)

The sources of the following circuits are contained in the Sources section, which begins on page 175. The figure number in the box of each circuit correlates to the source entry in the Sources section.

Universal State-Variable Filter
Three-Amplifier Active Filter
State Variable Filter
Second-Order State-Variable Filter (1 kHz, $Q = 10$)
Biquad Filter

Digitally Tuned Low-Power Active Filter
Tunable Active Filter
Active RC Filter for Frequencies up to 150 kHz
5-Pole Active Filter
State Variable Filter with Multiple Filtering Outputs

UNIVERSAL STATE-VARIABLE FILTER

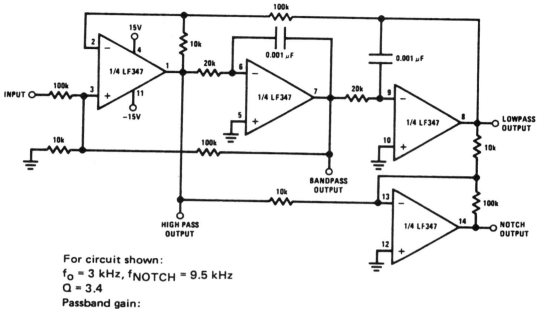

For circuit shown:

f_O = 3 kHz, f_{NOTCH} = 9.5 kHz

Q = 3.4

Passband gain:

    Highpass — 0.1

    Bandpass — 1

    Lowpass — 1

    Notch — 10

- f_O x Q $\leq$ 200 kHz
- 10V peak sinusoidal output swing without slew limiting to 200 kHz
- See LM348 data sheet for design equations

NATIONAL SEMICONDUCTOR

Fig. 10-1

STATE-VARIABLE ACTIVE FILTER

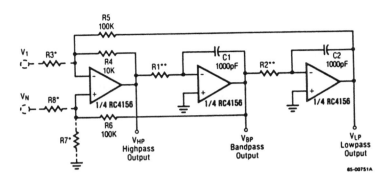

*Input connections are chosen for inverting or
non-inverting response. Values of R3, R7, R8
determine gain and Q.
**Values of R1 and R2 determine natural
frequency.

RAYTHEON

Fig. 10-2

A generalized circuit diagram of the two-pole state-variable active filter is shown. The state-variable filter can be inverting or noninverting and can simultaneously provide three outputs: low-pass, bandpass, and high-pass. A notch filter can be realized by adding one summing op amp.

In the state-variable filter circuit, one amplifier performs a summing function and the other two act as integrators. The choice of passive component values is arbitrary, but must be consistent with the amplifier operating range and input signal characteristics. The values shown for C1, C2, R4, R5, and R6 are arbitrary. Preselecting their values will simplify the filter tuning procedures, but other values can be used if necessary.

THREE-AMPLIFIER ACTIVE FILTER

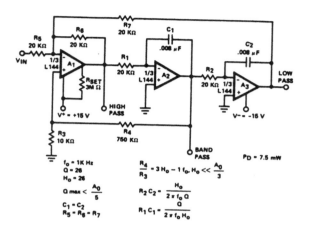

$f_o = 1K\ Hz$
$Q = 26$
$H_o = 26$

$Q_{max} < \dfrac{A_0}{5}$

$C_1 = C_2$
$R_5 = R_8 = R_7$

$\dfrac{R_4}{R_3} = 3\,H_o - 1\,f_o,\ H_o \ll \dfrac{A_0}{3}$

$R_2\,C_2 = \dfrac{H_o}{2\,\pi\,f_o\,Q}$

$R_1\,C_1 = \dfrac{Q}{2\,\pi\,f_o\,H_o}$

$P_D = 7.5\ mW$

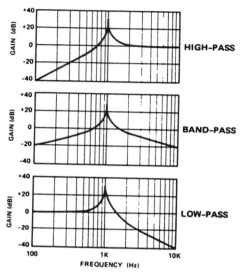

Bode plots of Active Filter Output

SILICONIX

Fig. 10-3

The active filter is a state-variable filter with bandpass, high-pass, and low-pass outputs. It uses a classic analog computer method of implementing a filter using three amplifiers and only two capacitors.

STATE VARIABLE FILTER

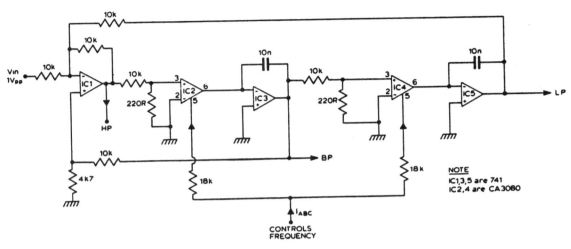

Fig. 10-4

This filter produces three outputs: high-pass, bandpass, and low-pass. Frequency is linearly proportional to the gain of the two integrators. Two CA3080s, (IC2 and IC4) provide the variable gain, the resonant frequency being proportional to the current, I_{ABC}. Using 741 op amps for IC3, a control range of 100 to 1 (resonant frequency) can be obtained. If CA3140s are used instead of 741s this range can be extended to nearly 10,000 to 1.

SECOND-ORDER STATE-VARIABLE FILTER (1 kHz Q = 10)

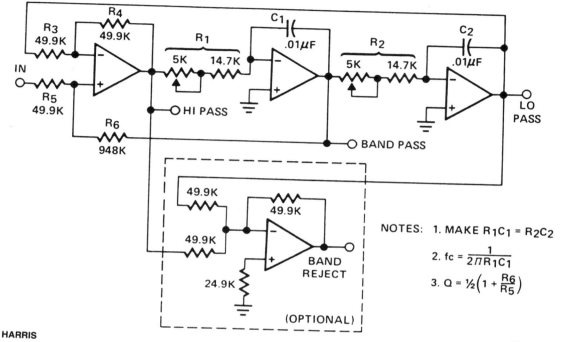

NOTES: 1. MAKE $R_1C_1 = R_2C_2$

2. $fc = \dfrac{1}{2\pi R_1 C_1}$

3. $Q = \frac{1}{2}\left(1 + \dfrac{R_6}{R_5}\right)$

HARRIS

Fig. 10-5

BIQUAD FILTER

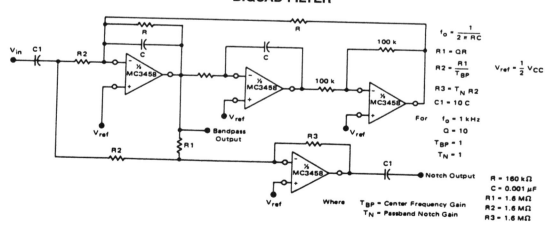

$f_o = \dfrac{1}{2\pi RC}$

$R1 = QR$

$R2 = \dfrac{R1}{T_{BP}}$ $V_{ref} = \frac{1}{2} V_{CC}$

$R3 = T_N R2$

$C1 = 10 C$

For $f_o = 1\,kHz$

$Q = 10$

$T_{BP} = 1$

$T_N = 1$

$R = 160\,k\Omega$
$C = 0.001\,\mu F$
$R1 = 1.6\,M\Omega$
$R2 = 1.6\,M\Omega$
$R3 = 1.6\,M\Omega$

Where T_{BP} = Center Frequency Gain
        T_N = Passband Notch Gain

MOTOROLA

Fig. 10-6

DIGITALLY TUNED LOW-POWER ACTIVE FILTER

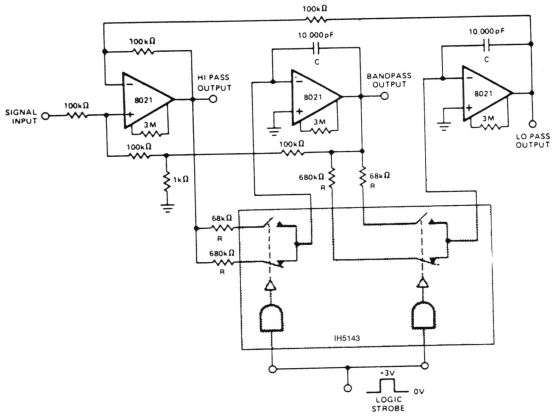

INTERSIL

Fig. 10-7

This constant-gain constant-Q, variable-frequency filter provides simultaneous low-pass, band-pass, and high-pass outputs. With the component values shown, the center frequency will be 235 Hz and 23.5 Hz for high and low logic inputs respectively, $Q=100$ and gain$=100$.

TUNABLE ACTIVE FILTER

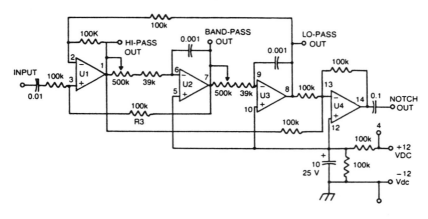

HAM RADIO

Fig. 10-8

The high-pass and low-pass outputs covering the range of 300 Hz to 3000 Hz have been summed in the fourth op amp to provide a notch output. The potentiometers must have a reverse log taper. The fixed-frequency active-filter center frequency is 1 kHz and the Q is 50.

ACTIVE RC FILTER FOR FREQUENCIES UP TO 150 kHz

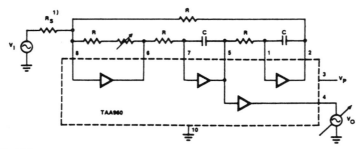

R = 10kΩ

This frequency range can be extended to 200kHz if a feed forward capacitor is connected between pin 5 and 8.

f	Frequency	$\frac{1}{2\pi RC}$	V
V_P	Supply voltage	6	V
	Filter performance		
Q	at $T_A = 25°C$	40 to 55	
Q	at $T_A = -30$ to $+65°C$	35 to 55	
V_i	Input voltage	400	mV
V_o	Output voltage	400	mV
d_{tot}	Distortion at $V_o = 350mV$	2	%
S/N	S/N ratio at $V_o = 400mV$	50	dB
R_s	Input resistor*	470	kΩ

SIGNETICS

*NOTE

Value of input resistor to be determined for $\frac{V_o}{V_i}$ = 0.90 to 1.1.

Fig. 10-9

5-POLE ACTIVE FILTER

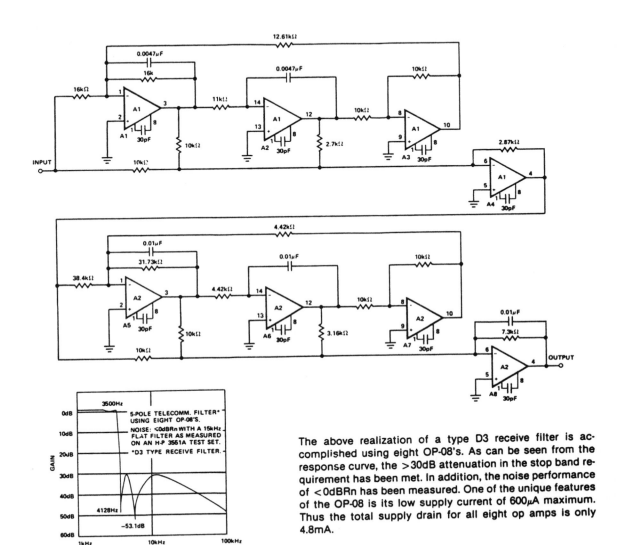

The above realization of a type D3 receive filter is accomplished using eight OP-08's. As can be seen from the response curve, the >30dB attenuation in the stop band requirement has been met. In addition, the noise performance of <0dBRn has been measured. One of the unique features of the OP-08 is its low supply current of 600µA maximum. Thus the total supply drain for all eight op amps is only 4.8mA.

PRECISION MONOLITHICS

Fig. 10-10

STATE-VARIABLE FILTER WITH MULTIPLE FILTERING OUTPUTS

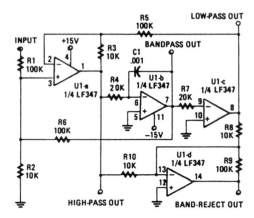

POPULAR ELECTRONICS   *Fig. 10-11*

11

Frequency Converters

The sources of the following circuits are contained in the Sources section, which begins on page 175. The figure number in the box of each circuit correlates to the source entry in the Sources section.

VLF Converter
Shortwave Converter
4- to 18-MHz Converter

Low-Frequency Converter
Simple LF Converter
100-MHz Converter

VLF CONVERTER

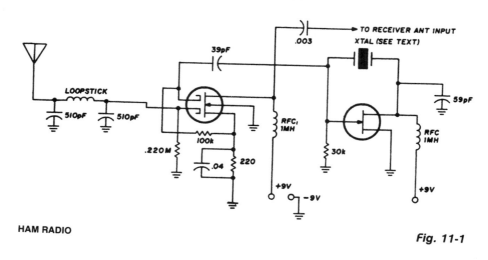

Fig. 11-1

This converter uses a low-pass filter instead of the usual tuned circuit so that the only tuning required is with the receiver. The dual-gate MOSFET and FET used in the mixer and oscillator aren't critical. Any crystal having a frequency compatible with the receiver tuning range can be used. For example, with a 3500-kHz crystal, 3500 kHz on the receiver dial corresponds to zero kHz; 3600 to 100 kHz; 3700 to 200 kHz, etc. (at 3500 kHz on the receiver, all you can hear is the converter oscillator; VLF signals start to come in about 20 kHz higher).

SHORTWAVE CONVERTER

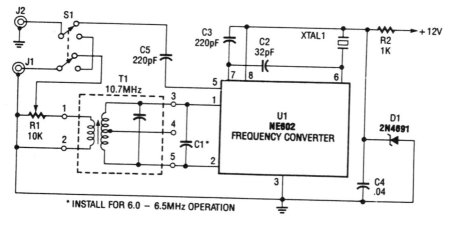

Fig. 11-2

SHORTWAVE CONVERTER *(Cont.)*

The NE602, U1, contains oscillator and mixer stages. The mixer combines the oscillator signal with the input RF signal to produce signals whose frequencies are the sum and difference of the input frequencies. For example, a 7.5-MHz signal is picked up by the antenna and mixes with the 8.5-MHz oscillator frequency. The difference between those two signals is 1 MHz—right in the center of the AM dial. Transformer T1 is a 10.7-MHz IF transformer.

4- TO 18-MHz CONVERTER

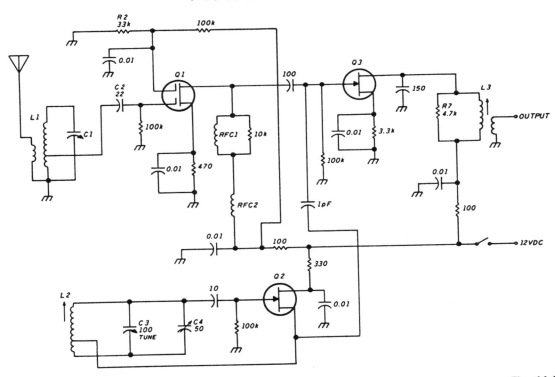

HAM RADIO

Fig. 11-3

The unit consists of RF amplifier Q1, local oscillator Q2, and mixer Q3. The two bands are covered without a bandswitch by using an IF of 3.5 MHz. The oscillator range is 7.5 to 14.5 MHz. Incoming signals from 4 to 11 MHz are mixed with the oscillator to produce the 3.5-MHz IF. Signals from 11 to 18 MHz are mixed with the oscillator to also produce an IF of 3.5 MHz. At any one oscillator frequency, the two incoming signals are 7 MHz apart. RF amplifier input C1/L1 consists of a high-Q lightly loaded tuned circuit; this is essential for good band separation.

LOW-FREQUENCY CONVERTER

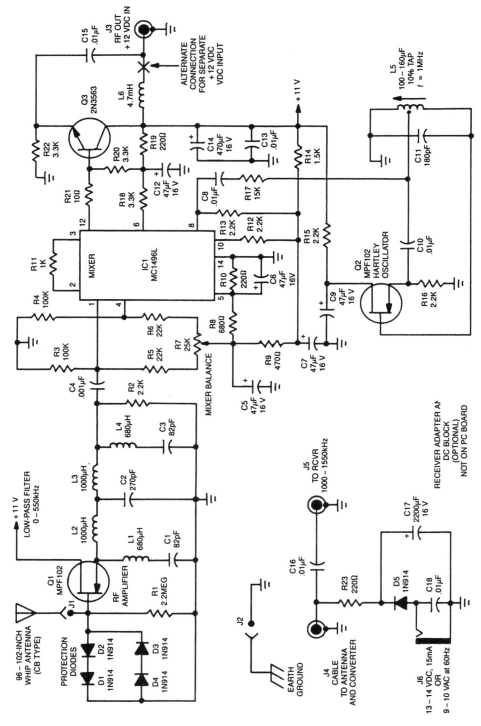

Fig. 11-4

Reprinted with permission from Radio-Electronics Magazine, September 1989. Copyright Gernsback Publications, Inc., 1989.

RADIO-ELECTRONICS

LOW-FREQUENCY CONVERTER *(Cont.)*

Among the signals below 550 kHz are maritime mobile, distress, radio beacon, aircraft weather, European longwave broadcast, and point-to-point communications. The low-frequency converter converts the 10- to 500-kHz LW range to a 1010- to 1500-kHz MW range, by adding 1000 kHz to all received signals. Radio calibration is unnecessary because signals are received at the AM-radio's dial setting, plus 1 MHz; a 100-kHz signal is received at 1100 kHz, a 335-kHz signal at 1335 kHz, etc. The low-frequency signals are fed to IC1, a doubly-balanced mixer.

Transistor Q2 and associated circuitry form a Hartley 1000-kHz local oscillator, which is coupled from Q2's drain, through C8, to IC1 pin 8. Signals in the 10- to 550-kHz range are converted to 1010 to 1550 kHz. The mixer heterodynes the incoming low-frequency signal and local-oscillator signal. Transistor Q3 reduces IC1's high-output impedance to about 100 Ω to match most receiver inputs. Capacitor C15 couples the 1010- to 1550-kHz frequencies from Q3's emitter to output jack J3 while blocking any dc bias.

Inductor L6 couples the dc voltage that's carried in the RF signal cable from the rcvr/dc adaptor. The dc voltage and RF signals don't interfere with one another; that saves running a separate power-supply wire, which simplifies installation at a remote location. Capacitors C14 and C13 provide dc supply filtering. The kit is available from North Country Radio, P.O. Box 53, Wykagyl Station, NY 10804.

SIMPLE LF CONVERTER

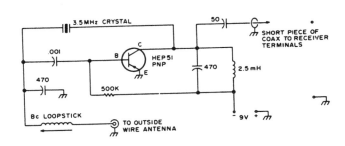

73 AMATEUR RADIO

Fig. 11-5

This converter allows coverage from 25 kHz up to 500 kHz. Use short coax from the converter to receiver antenna input. Tune the receiver to 3.5 MHz, peak for loudest crystal calibrator, and tune your receiver higher in frequency to 3.6 MHz. You're now tuning the 100-kHz range. 3.7 MHz puts you at 200 kHz, 3.8 MHz equals 300 kHz, 3.9 MHz yields 500 kHz, and 4.0 MHz gives you 500 kHz.

100-MHz CONVERTER

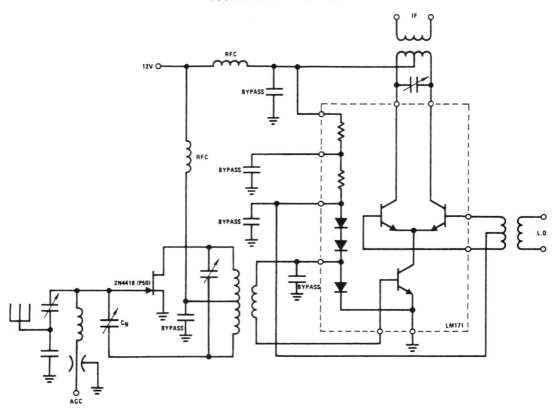

NATIONAL SEMICONDUCTOR

Fig. 11-6

The 2N4416 JFET will provide noise figures of less than 3 dB and power gain of greater than 20 dB. The JFET's outstanding low crossmodulation and low intermodulation distortion provides an ideal characteristic for an input stage. The output feeds into an LM171 used as a balanced mixer. This configuration greatly reduces local oscillator radiation both into the antenna and into the IF strip and also reduces RF signal feedthrough.

12

Frequency-to-Voltage Converters

The sources of the following circuits are contained in the Sources section, which begins on page 175. The figure number in the box of each circuit correlates to the source entry in the Sources section.

FREQUENCY-TO-VOLTAGE CONVERTERS

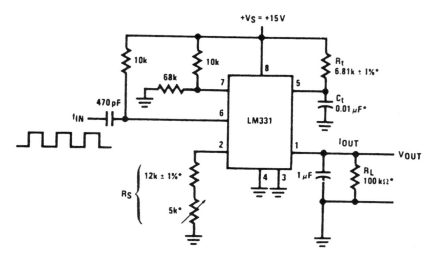

$$V_{OUT} = f_{IN} \times 2.09V \times \frac{R_L}{R_S} \times (R_tC_t)$$

*Use stable components with low temperature coefficients.

10 kHz Full-Scale, $\pm 0.006\%$ Nonlinearity

NATIONAL SEMICONDUCTOR

Fig. 12-1

In these applications, a pulse input at f_{IN} is differentiated by an RC network and the negative-going edge at pin 6 causes the input comparator to trigger the timer circuit. Just as with a V/F converter, the average current flowing out of pin 1 is $I_{AVERAGE} = i \times (1.1\,R_1C_1) \times f$. In this simple circuit, this current is filtered in the network $R_L = 100$ kΩ and 1 μF. The ripple will be less than 10 mV peak, but the response will be slow, with a 0.1-second time constant, and settling of 0.7 second to 0.1% accuracy. In the precision circuit, an operational amplifier provides a buffered output and also acts as a 2-pole filter. The ripple will be less than 5 mV peak for all frequencies above 1 kHz, and the response time will be much quicker than in Part 1. However, for input frequencies below 200 Hz, this circuit will have worse ripple than the figure. The engineering of the filter time constants to get adequate response and small enough ripple simply requires a study of the compromises to be made. Inherently, V/F converter response can be fast, but F/V response cannot.

FREQUENCY-TO-VOLTAGE CONVERTERS (Cont.)

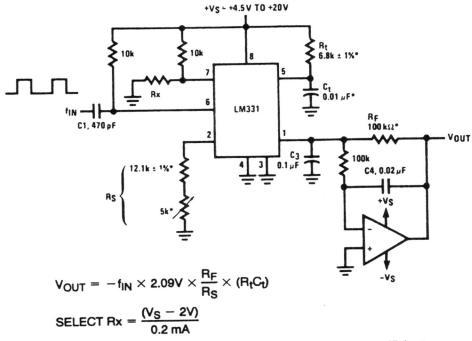

$$V_{OUT} = -f_{IN} \times 2.09V \times \frac{R_F}{R_S} \times (R_t C_t)$$

$$\text{SELECT } R_x = \frac{(V_S - 2V)}{0.2 \text{ mA}}$$

*Use stable components with low temperature coefficients.

10 kHz Full-Scale with 2-Pole Filter, ±0.01% Nonlinearity Maximum

DC TO 10-kHz FREQUENCY-TO-VOLTAGE CONVERTER

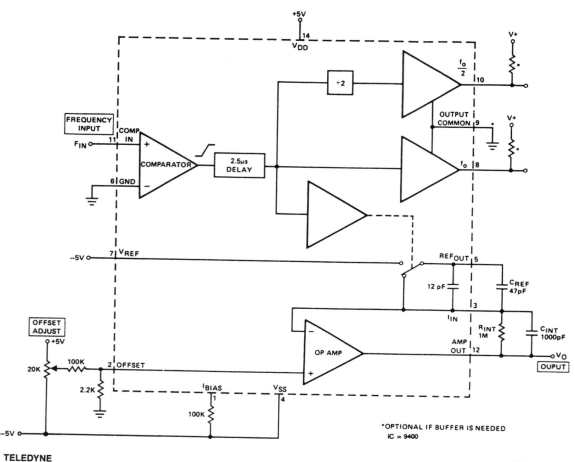

TELEDYNE

Fig. 12-2

The converter generates an output voltage, which is linearly proportional to the input frequency waveform. Each zero crossing at the comparator's input causes a precise amount of change to be dispensed into the op amp's summing junction. This charge, in turn, flows through the feedback resistor generating voltage pulses at the output of the op amp. Capacitor (C_{INT}) across R_{INT} averages these pulses into a dc voltage, which is linearly proportional to the input frequency.

FREQUENCY-TO-VOLTAGE CONVERTER
(DIGITAL FREQUENCY METER)

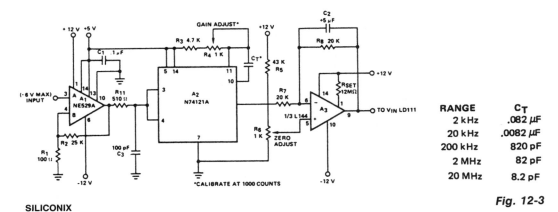

RANGE	C_T
2 kHz	.082 μF
20 kHz	.0082 μF
200 kHz	820 pF
2 MHz	82 pF
20 MHz	8.2 pF

SILICONIX

Fig. 12-3

This circuit converts frequency to voltage by taking the average dc value of the pulses from the 74121 monostable multivibrator. The one shot is triggered by the positive-going ac signal at the input of the 529 comparator. The amplifier acts as a dc filter, and also provides zeroing. The accuracy is 2% over a five-decade range. The input signal to the comparator should be greater than 0.1 V peak-to-peak, and less than 12 V peak-to-peak for proper operation.

ZENER-REGULATED FREQUENCY-TO-VOLTAGE CONVERTER

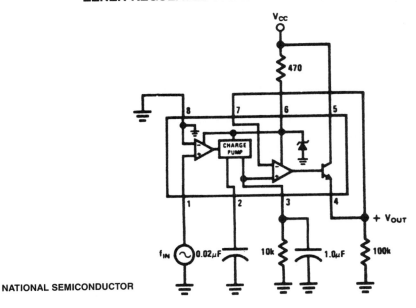

NATIONAL SEMICONDUCTOR

Fig. 12-4

109

FREQUENCY-TO-VOLTAGE
CONVERTER WITH TTL INPUT

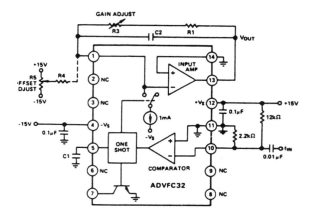

ANALOG DEVICES

Fig. 12-5

FREQUENCY-TO-VOLTAGE CONVERTER WITH 2-POLE
BUTTERWORTH FILTER TO REDUCE RIPPLE

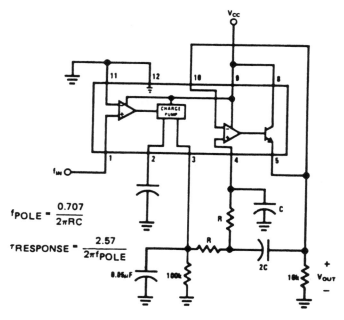

$$f_{POLE} = \frac{0.707}{2\pi RC}$$

$$\tau_{RESPONSE} = \frac{2.57}{2\pi f_{POLE}}$$

NATIONAL SEMICONDUCTOR

Fig. 12-6

FREQUENCY-TO-VOLTAGE CONVERTER

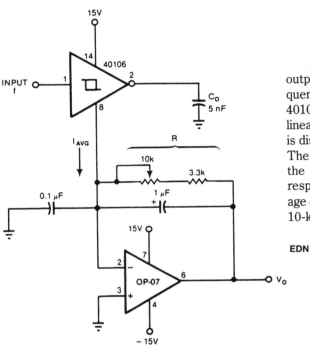

Six components can configure a circuit whose output voltage is proportional to its input frequency. The average current (I_{AVG}) from the 40106 Schmitt trigger inverter's ground pin 8 is linearly dependent on the frequency at which $C0$ is discharged into the op amp's summing junction. The op amp forces this current to flow through the 13.33-kΩ feedback resistor, producing a corresponding voltage drop. This frequency-to-voltage converter yields 0- to -10-V output with 0- to 10-kHz input frequencies.

EDN

Fig. 12-7

13

Miscellaneous Converters

The sources of the following circuits are contained in the Sources section, which begins on page 175. The figure number in the box of each circuit correlates to the source entry in the Sources section.

PICOAMPERE-TO-FREQUENCY CONVERTERS

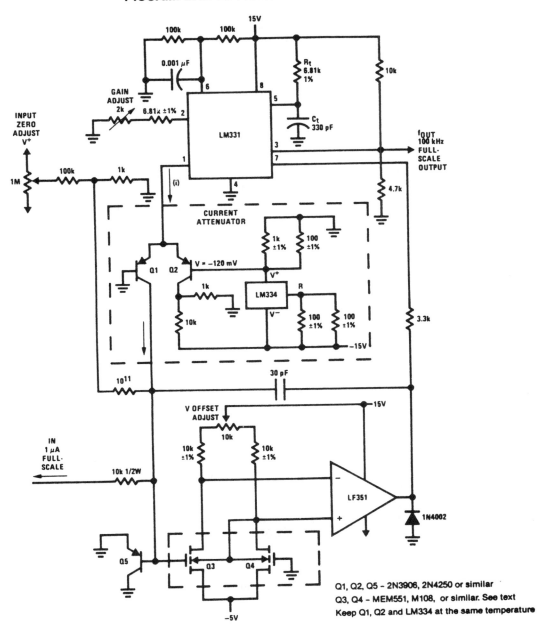

Q1, Q2, Q5 – 2N3906, 2N4250 or similar
Q3, Q4 – MEM551, M108, or similar. See text
Keep Q1, Q2 and LM334 at the same temperature

Fig. 13-1

2 SIMPLE TEMPERATURE-TO-TIME CONVERTERS

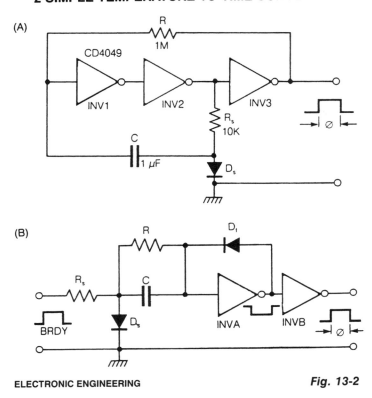

ELECTRONIC ENGINEERING

Fig. 13-2

Both of these converters use CMOS inverters. Figure 13-2A shows a free-running circuit that has both the pulse duration and pulse pause dependent on temperature of the diode, D_S. It can be used where a synchronization between the converter and something else is not required.

Figure 13-2B shows one-shot circuit that produces a pulse with its duration dependent of temperature of diode D_s. The additional diode D_f should have inverse current low enough to not influence the discharging process in the RC network when the INVA output is low. A silicon component or a GaAs LED can be used.

The converter is intended for a digital system producing a RADY pulse, which disappears after the process has ended. The pulse duration is approximately:

$$= 2RC\frac{V_D}{V_{DD}}$$

where V_D is the sensor diode forward voltage and V_{DD} is the supply voltage of the CMOS chip.

Resistance R must be much higher than R_S. A 0.1-μF capacitor can be applied in parallel with D_S, if necessary, to repulse stray pickup and noise in a long cable. The circuits described can be used with a temperature-sensitive resistor replacing diode D_S.

VOLTAGE RATIO-TO-FREQUENCY CONVERTER

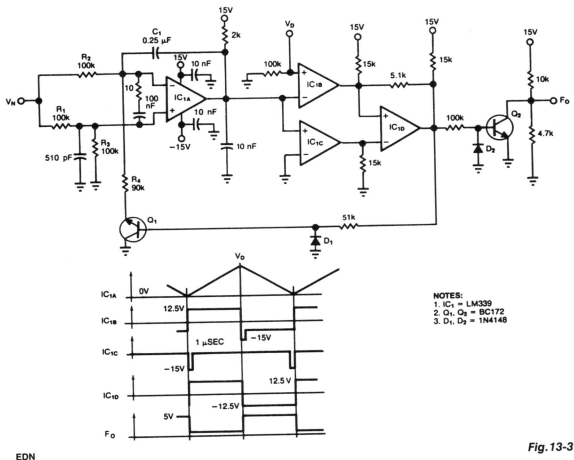

NOTES:
1. IC_1 = LM339
2. Q_1, Q_2 = BC172
3. D_1, D_2 = 1N4148

Fig. 13-3

EDN

The circuit accepts two positive-voltage inputs V_N and V_D and provides a TTL-compatible output pulse train whose repetition rate is proportional to the ratio V_N/V_D. Full-scale output frequency is about 100 Hz, and linearity error is below 0.5 percent. The output F_o equals KV_N/V_D, where $K = 1/(4R_2C_1)$ provided that $R_1 = R_3$. Op amp IC1A alternately integrates $V_N/2$ and $-V_N/2$, producing a sawtooth output that ramps between the V_D level and ground. When transistor Q1 is on, for example, IC1A integrates $-V_N/2$ until its output equals V_D. At that time, the IC1B comparator switches low, causing IC1D's bistable output to go low, which turns off Q1. IC1A's output then ramps in the negative direction. When the output reaches 0 V, the IC1C comparator switches, Q1 turns on, and the cycle repeats. Transistor Q2 converts the IC1D output to TTL-compatible output logic levels. Setting V_D to 1.00 V yields a linear voltage-to-frequency converter ($F_o = KV_D$).

50-MHz THERMAL RMS-TO-DC CONVERTER

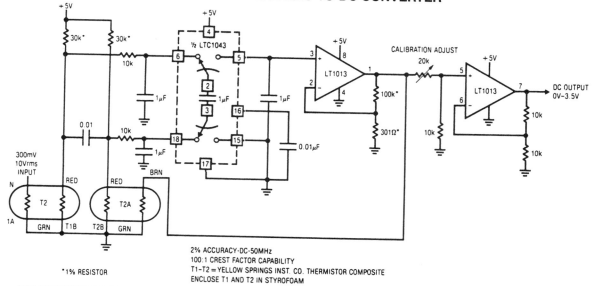

2% ACCURACY-DC-50MHz
100:1 CREST FACTOR CAPABILITY
T1-T2 = YELLOW SPRINGS INST. CO. THERMISTOR COMPOSITE
ENCLOSE T1 AND T2 IN STYROFOAM

*1% RESISTOR

LINEAR TECHNOLOGY

Fig. 13-4

PULSE WIDTH-TO-VOLTAGE CONVERTER

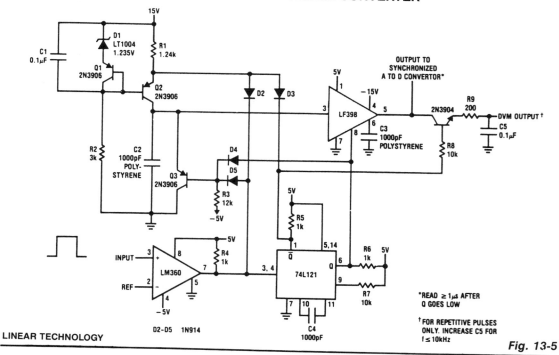

*READ ≥ 1µs AFTER
Q GOES LOW

† FOR REPETITIVE PULSES
ONLY. INCREASE C5 FOR
f ≤ 10kHz

D2-D5 1N914

LINEAR TECHNOLOGY

Fig. 13-5

PIN PHOTODIODE-TO-FREQUENCY CONVERTER

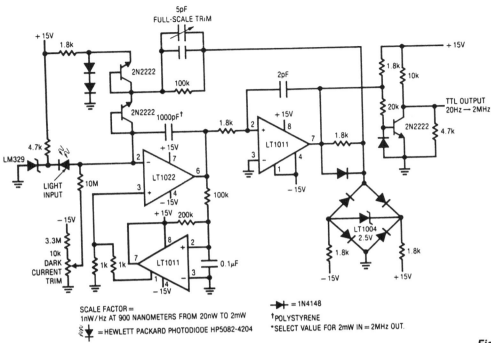

SCALE FACTOR =
1nW/Hz AT 900 NANOMETERS FROM 20nW TO 2mW

↯⟩ = HEWLETT PACKARD PHOTODIODE HP5082-4204

─▶⊢ = 1N4148

†POLYSTYRENE

*SELECT VALUE FOR 2mW IN = 2MHz OUT.

LINEAR TECHNOLOGY

Fig. 13-6

0 I_B ERROR V/I CONVERTER

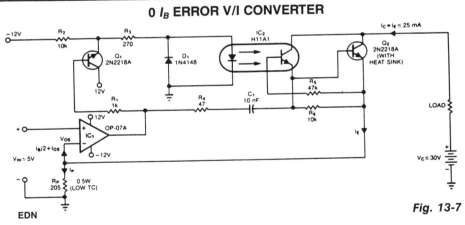

EDN

Fig. 13-7

Single programming resistor R_p provides an output-current range of about six decades. Notice that this resistor's TC is also a potential source of error; it dissipates 125 mW when V_{IN} = 5 V. The maximum deviation is typically 50 nA or 0.0002% of full scale. This voltage-controlled current source uses optocoupler IC2 to eliminate an error found in more conventional circuits, which is caused by the output transistor's base current.

117

TRIANGLE-TO-SINE CONVERTERS

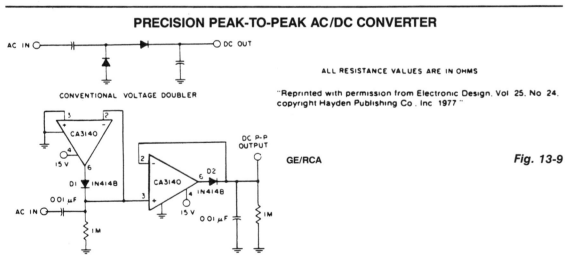

SIGNETICS

Fig. 13-8

Conversion of triangle waves to sinusoids is usually accomplished by diode/resistor shaping networks, which accurately reconstructs the sine wave, segment by segment. Two simpler and less costly methods may be used to shape the triangle waveform of the 566 into a sinusoid with less than 2% distortion. The non-linear $I_{DS}V_{DS}$ transfer characteristic of a p-channel junction FET is used to shape the triangle waveform. The amplitude of the triangle waveform is critical and must be carefully adjusted to achieve a low-distortion sinusoidal output. Naturally, where additional waveform accuracy is needed, the diode-resistor shaping scheme can be applied to the 566 with excellent results because it has very good output amplitude stability when operated from a regulated supply.

PRECISION PEAK-TO-PEAK AC/DC CONVERTER

ALL RESISTANCE VALUES ARE IN OHMS

GE/RCA

Fig. 13-9

A CA3140 BiMOS op amp and a single positive supply converts a conventional voltage doubler with two precision diodes into a precision peak-to-peak ac-to-dc voltage converter that has wide dynamic range and wide bandwidth.

118

PHOTODIODE CURRENT-TO-VOLTAGE CONVERTER

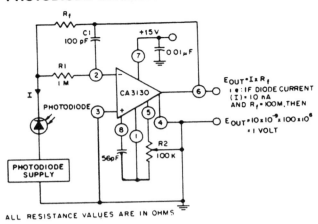

$$E_{OUT} = I \times R_f$$
I • : IF DIODE CURRENT
(I) = 10 nA
AND R_f = 100M, THEN

$$E_{OUT} = 10 \times 10^{-9} \times 100 \times 10^{6}$$
= 1 VOLT

ALL RESISTANCE VALUES ARE IN OHMS

GE/RCA

Fig. 13-10

The circuit uses three CA3130 BiMOS op amps in an application sensitive to subpicoampere input currents. The circuit provides a ground-referenced output voltage proportional to input current flowing through the photodiode.

SELF-OSCILLATING FLYBACK CONVERTER

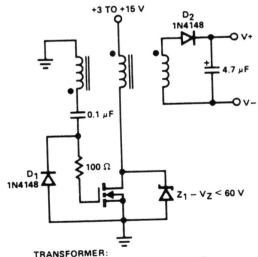

TRANSFORMER:
INDIANA GENERAL CORE F626-12-Q_2
26 TURNS NO. 28 WIRE TRIFILAR WOUND

This low-power converter uses the core characteristics to determine frequency. With the transformer shown, the operating frequency is 250 kHz. Diode D1 prevents negative spikes from occuring at the MOSFET gate, the 100-Ω resistor is a parasitic suppressor, and Z1 serves as a dissipative voltage regulator for the output and also clips the drain voltage to a level below the rated power FET breakdown voltage.

SILICONIX

Fig. 13-11

BCD-TO-ANALOG CONVERTER

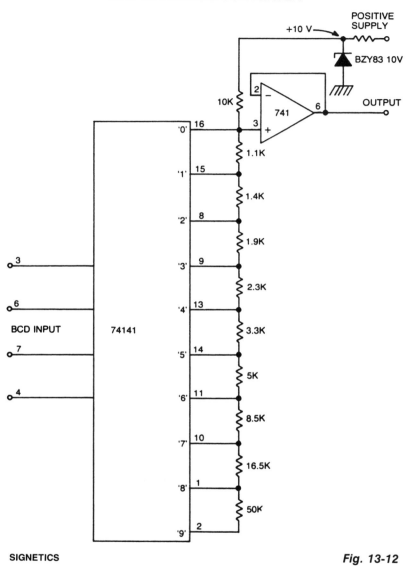

SIGNETICS

Fig. 13-12

This circuit will convert four-bit BCD into a variable voltage from 0 to 9 V in 1-V steps. The SN74141 is a nixie driver, and has 10 open-collector outputs. These are used to ground a selected point in the divider chain which is determined by the BCD code at the input, and so produce a corresponding voltage at the output. Accuracy of the circuit depends on the tolerance of the resistors and the accuracy of the reference voltage. However, presets can be used in the divider chain, with correct calibration. The 741 is used as a buffer.

RESISTANCE-TO-VOLTAGE CONVERTER

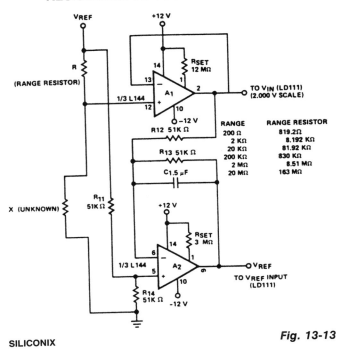

Fig. 13-13

This circuit will measure accurately to 20 MΩ when associated with a buffer amplifier (A1), which has a low input bias current (I_{IN}) < 30nA). The circuit uses two of the three amplifiers contained in the Siliconix L144 micropower triple op amp.

LOW-COST μP-INTERFACED TEMPERATURE-TO-DIGITAL CONVERTER

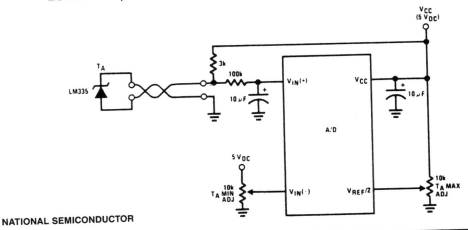

Fig. 13-14

HIGH-IMPEDANCE PRECISION RECTIFIER FOR AC/DC CONVERTER

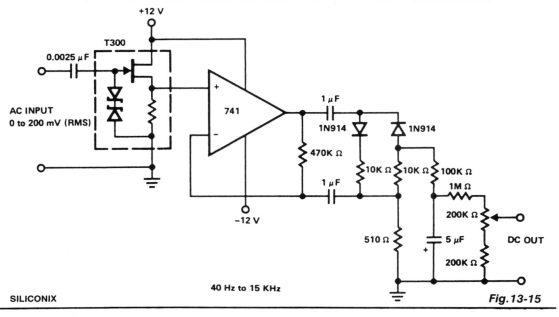

Fig. 13-15

WIDE-RANGE CURRENT-TO-FREQUENCY CONVERTER

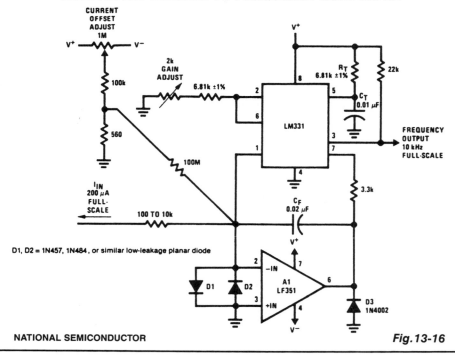

D1, D2 = 1N457, 1N484, or similar low-leakage planar diode

NATIONAL SEMICONDUCTOR

Fig. 13-16

WIDEBAND HIGH-CREST FACTOR RMS-TO-DC CONVERTER

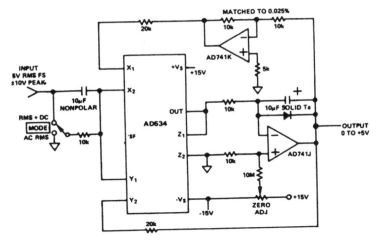

CALIBRATION PROCEDURE:

WITH 'MODE' SWITCH IN 'RMS + DC' POSITION, APPLY AN INPUT OF + 1.00VDC. ADJUST ZERO UNTIL OUTPUT READS SAME AS INPUT. CHECK FOR INPUTS OF ± 0.06% (5mV).

ACCURACY IS MAINTAINED FROM 60Hz TO 100kHz, AND IS TYPICALLY HIGH BY 0.6% AT 1MHz FOR V_{IN}-4V RMS (SINE, SQUARE, OR TRIANGULAR WAVE).

PROVIDED THAT THE PEAK INPUT IS NOT EXCEEDED, CREST-FACTORS UP TO AT LEAST TEN HAVE NO APPRECIABLE EFFECT ON ACCURACY.

INPUT IMPEDANCE IS ABOUT 10kΩ; FOR HIGH (10MΩ) IMPEDANCE, REMOVE MODE SWITCH AND INPUT COUPLING COMPONENTS.

FOR GUARANTEED SPECIFICATIONS THE AD536A AND AD636 IS OFFERED AS A SINGLE PACKAGE RMS-TO-DC CONVERTER.

ANALOG DEVICES

Fig. 13-17

LIGHT INTENSITY-TO-FREQUENCY CONVERTER

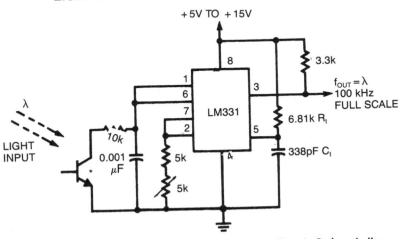

*L14F-1, L14G-1 or L14H-1, photo transistor (General Electric Co.) or similar

NATIONAL SEMICONDUCTOR

Fig. 13-18

MULTIPLEXED BCD-TO-PARALLEL BCD CONVERTER

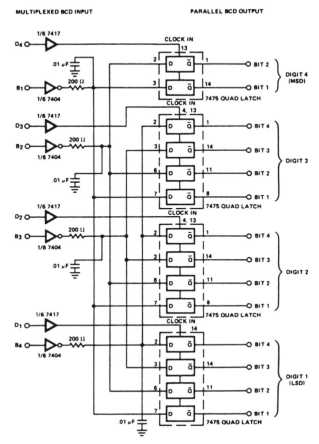

SILICONIX

Fig. 13-19

This converter consists of four quad bistable latches activated in the proper sequence by the digit strobe output of the LD110. The complemented outputs (Q) of the quad latch set reflect the state of the bit outputs when the digit strobe goes high. It will maintain this state when the digit strobe goes low.

FAST LOGARITHMIC CONVERTER

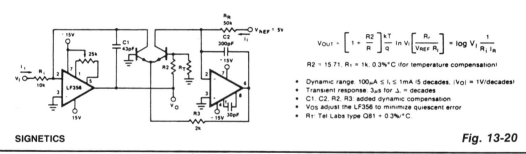

$$V_{OUT} = \left[1 + \frac{R2}{R}\right]\frac{kT}{q}\ln V_I\left[\frac{R_r}{V_{REF}\,R_I}\right] = \log V_I \frac{1}{R_I\,I_R}$$

R2 = 15.71, R_I = 1k, 0.3%°C (for temperature compensation)

- Dynamic range: 100μA ≤ I, ≤ 1mA (5 decades, |Vo| = 1V/decades)
- Transient response: 3μs for Δ = decades
- C1, C2, R2, R3: added dynamic compensation
- Vos adjust the LF356 to minimize quiescent error
- R⊤: Tel Labs type Q81 + 0.3%/°C.

SIGNETICS

Fig. 13-20

SINE-WAVE/SQUARE-WAVE CONVERTER

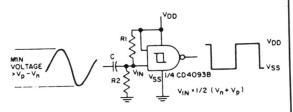

RCA

Fig. 13-21

The sine input is ac coupled by capacitor C; R1 and R2 bias the input midway between V_n and V_p, the input threshold voltages, to provide a square wave at the output.

TTL-TO-MOS LOGIC CONVERTER

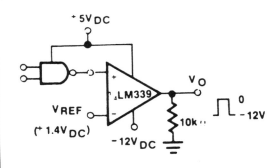

SIGNETICS

Fig. 13-23

SELF-OSCILLATING FLYBACK CONVERTER

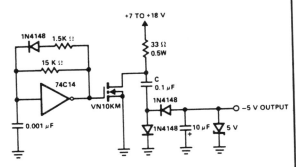

SILICONIX

Fig. 13-22

This low-power converter is suitable for deriving a higher voltage from a main system rail in an on board application. With the transformer shown, the operating frequency is 250 kHz. Z1 serves as a dissipative voltage regulator for the output and also clips the drain voltage to a level below the rated VMOS breakdown voltage.

PICOAMPERE-TO-VOLTAGE CONVERTER WITH GAIN

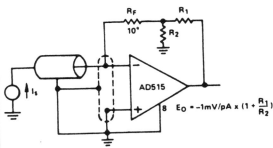

ANALOG DEVICES

Fig. 13-24

COMPLEMENTARY-OUTPUT VARIABLE-FREQUENCY INVERTER

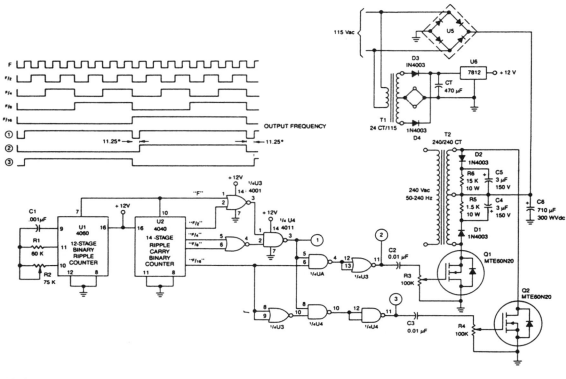

MOTOROLA

Fig. 13-25

U1 is a 4060 12-stage binary ripple counter that is used as a free-running oscillater; its frequency of oscillation is: $1/2.2\ C1R2$. The output of U1 is applied to U2, a 14-stage binary ripple counter that provides square-wave outputs of $1/2$, $1/4$, $1/8$, and $1/16$ of the clock frequency. These signals are combined in U3 and U4 to provide a complementary drive for Q1 and Q2.

Outputs from U3 and U4 are ac-coupled to Q1 and Q2 via C2 and C4, respectively. R3 and R4 adjust the gate drive to Q1 and Q2. Q1 and Q2 alternately draw current through opposing sides of the primary to synthesize an ac input voltage at a given frequency. Only one side of the primary of T2 is driven at one time, so maximum power output is half of the transformer rating.

PULSE HEIGHT-TO-WIDTH CONVERTER

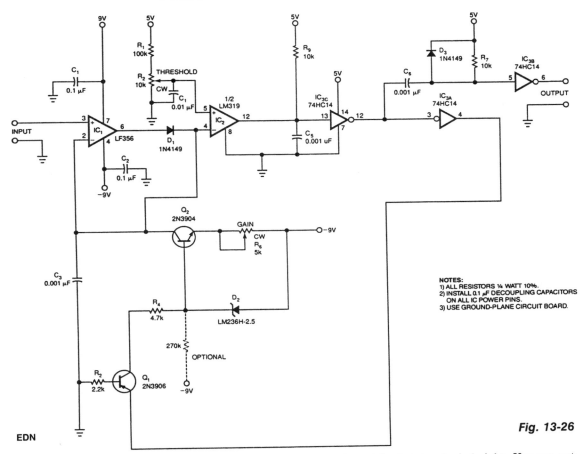

NOTES:
1) ALL RESISTORS ¼ WATT 10%.
2) INSTALL 0.1 μF DECOUPLING CAPACITORS ON ALL IC POWER PINS.
3) USE GROUND-PLANE CIRCUIT BOARD.

EDN

Fig. 13-26

The output-pulse width from the circuit is a linear function of the input pulse's height. You can set the circuit's input threshold to discriminate against low-level pulses, while fixed components limit the circuit's maximum output-pulse width.

With a 270-kΩ resistor connected from the −9V supply to the base lead of Q2, this circuit can handle input pulses separated by 20 μs for correct operation. The turn-off timer of zener-diode D2 sets the lower limit for input-pulse repetition rate.

IC1, D1 and C3 detect the peak of the input pulse. The computer IC2, triggers at your preset threshold. The RC delay network, R9 and C5, hold off inverter IC3's changing state until the completion of peak detection. After IC3A changes state, Q1 turns on and then turns on Q2, a constant-current source.

Constant-current source Q1 then discharges C3, the peak-detecting capacitor. When C3 has discharged below IC2's threshold, IC2's output decreases, as do pins 6 and 4 of IC3. The output-pulse width is a function of this discharge time, which you can adjust with R6. C6 and R7 control the maximum output pulse width, which is 8 μs max.

PULSE TRAIN-TO-SINUSOID CONVERTER

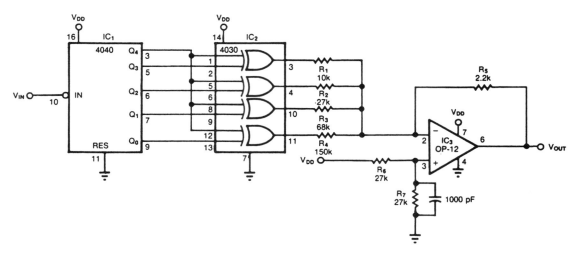

Fig. 13-27

This circuit lets you convert a serial pulse stream or sinusoidal input to a sinusoidal output at $1/32$ of the frequency. By varying the frequency of V_{IN}, you can achieve an output range of $10^7:1$—from about 100 kHz to less than 0.01 Hz. The output resembles that of a 5-bit D/A converter operating on parallel digital data.

Counter IC1 generates binary codes that repeatedly scan the range from 00000 to 11111. The output amplifier adds the corresponding XOR gate outputs, V_{DD} or ground, weighted by the values of input resistors R1 through R4. The 16 counter codes 00000 to 01111, for instance, pass unchanged to the XOR gate outputs, and cause V_{OUT} to step through the half-sinusoidal cycle for maximum amplitude to minimum amplitude.

Counter output Q4 becomes high for the next 16 codes and causes the XOR gates to invert the Q0 through Q3 outputs. As a result, V_{OUT} steps through the remaining half cycle from minimum to maximum amplitude. The counter then rolls over and initiates the next cycle. You can change the R1 through R4 values to obtain other V_{OUT} waveforms. V_{DD} should be at least 12 V to assure maximum-frequency operation from IC1 to IC2.

OHMS-TO-VOLTS CONVERTER

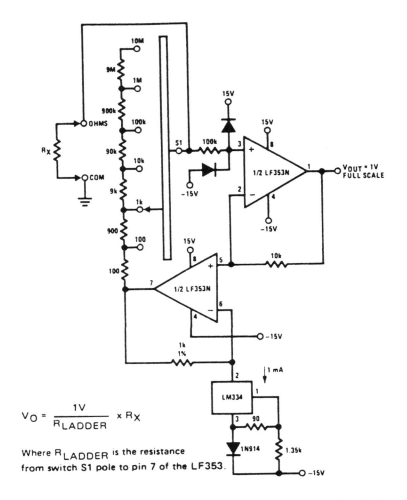

$$V_O = \frac{1V}{R_{LADDER}} \times R_X$$

Where R_{LADDER} is the resistance from switch S1 pole to pin 7 of the LF353.

NATIONAL SEMICONDUCTOR

Fig. 13-28

129

TTL SQUARE WAVE-TO-TRIANGLE CONVERTER

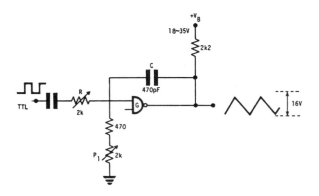

ELECTRONIC ENGINEERING

Fig. 13-29

This fixed frequency triangular waveform generator driven by a TTL square wave generates typically 16-V p-p triangles at frequencies up to several MHz. It uses only one NAND open-collector gate, or one open collector inverter as a fast integrator with gain. Careful successive adjustments of R and P1 are needed. When correct adjustments are reached, output amplitude and linearity are largely independent of the value of V_B, from a minimum of 18 V up to 35 V.

SQUARE-TO-SINE WAVE CONVERTER

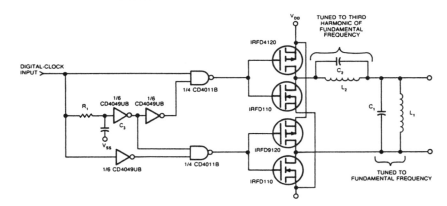

EDN

Fig. 13-30

Two pairs of MOSFETs form a bridge that alternately switches current in opposite directions. Two parallel-resonant LC circuits complete the converter. The L1/C1 combination is resonant at the fundamental frequency; the L2/C2 combination is resonant at the clock frequency's third harmonic and acts as a trap. T1and C3 ensure that both halves of the MOSFET bridge are never on at the same time by providing a common delay to the gate drive of each half.

CAPACITANCE-TO-PULSE WIDTH CONVERTER

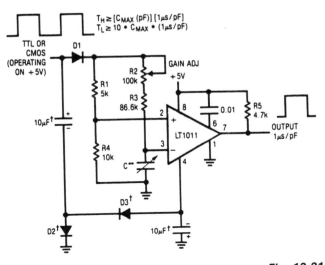

$T_H \geq [C_{MAX} (pF)] [1\mu s / pF]$
$T_L \geq 10 \cdot C_{MAX} \cdot (1\mu s / pF)$

TTL OR CMOS (OPERATING ON +5V)

GAIN ADJ

R1 5k
R2 100k
R3 86.6k
R4 10k
R5 4.7k

0.01

LT1011

OUTPUT $1\mu s / pF$

$10\mu F^\dagger$

C^{**}

D1

$D3^\dagger$

$D2^\dagger$

$10\mu F^\dagger$

*$PW = (R2 + R3) (C) \left(\dfrac{R1 + R4}{R1} \right)$, INPUT CAPACITANCE OF
LT1011 IS ≈6pF. THIS IS AN OFFSET TERM.
†THESE COMPONENTS MAY BE ELIMINATED IF NEGATIVE SUPPLY IS AVAILABLE (−1V TO −15V).
**TYPICAL 2 SECTIONS OF 365pF VARIABLE CAPACITOR WHEN USED AS SHAFT ANGLE INDICATION.

LINEAR TECHNOLOGY

Fig. 13-31

HIGH/LOW RESISTANCE-TO-VOLTAGE CONVERTER

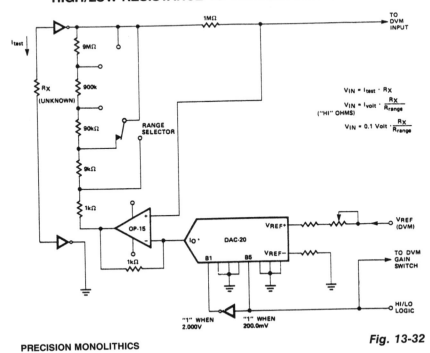

I_{test}

1MΩ

TO DVM INPUT

9MΩ

RX (UNKNOWN)

900k

90kΩ

RANGE SELECTOR

9kΩ

1kΩ

OP-15

1kΩ

Io

DAC-20

VREF+

VREF−

B1   B5

VREF (DVM)

TO DVM GAIN SWITCH

HI/LO LOGIC

"1" WHEN 2.000V

"1" WHEN 200.0mV

$V_{IN} = I_{test} \cdot R_X$
$V_{IN} = I_{volt} \cdot \dfrac{R_X}{R_{range}}$ ("HI" OHMS)
$V_{IN} = 0.1 \text{ Volt} \cdot \dfrac{R_X}{R_{range}}$

PRECISION MONOLITHICS

Fig. 13-32

131

VOLTAGE-TO-PULSE DURATION CONVERTER

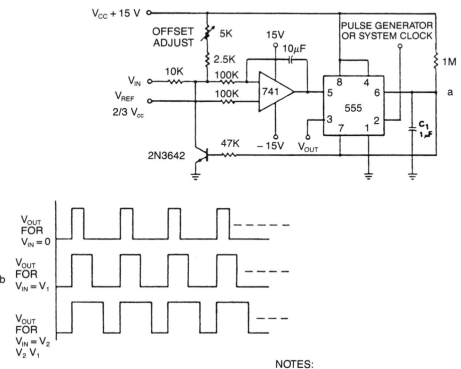

NOTES:
All resistor values in ohms
*V_{IN} is limited to 2 diode drops within ground or below V_{CC}

Fig. 13-33

Voltage levels can be converted to pulse durations by combining an op amp and a timer IC. Accuracies to better than 1% can be obtained with this circuit (a), and the output signals (b) still retain the original frequency, independent of the input voltage.

AC-TO-DC CONVERTER

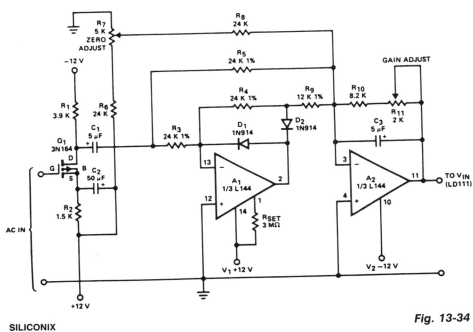

SILICONIX

Fig. 13-34

The circuit includes a PMOS enhancement-mode FET input buffer amplifier, coupled to a classic absolute-value circuit, which essentially eliminates the effect of the forward-voltage drop across diodes D1 and D2.

POLARITY CONVERTER

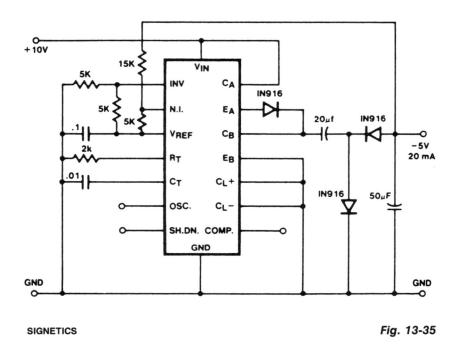

SIGNETICS

Fig. 13-35

The capacitor diode output circuit is used here as a polarity converter to generate a −5-V supply from +15 V. This circuit is useful for an output current of up to 20 mA with no additional boost transistors required. Because the output transistors are current limited, no additional protection is necessary. Also, the lack of inductor allows the circuit to be stabilized with only the output capacitor.

CALCULATOR-TO-STOPWATCH CONVERTER

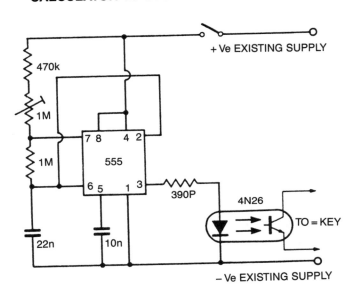

ELECTRONICS TODAY INTERNATIONAL

Fig. 13-36

This circuit can be fitted to any calculator with an automatic constant to enable it to be used as a stopwatch. The 555 timer is set to run at a suitable frequency and connected to the existing calculator battery via the push-on push-off switch and the existing calculator on-off switch.

14

Power Converters/Inverters

The sources of the following circuits are contained in the Sources section, which begins on page 175. The figure number in the box of each circuit correlates to the source entry in the Sources section.

dc/dc Regulating Converter
Flyback Converter
1.5-W Offline Converter
Dual-Output ±12 or ±15V dc/dc Converter
12-V Converter
Regulated 6-V to 15-V Flyback Converter
Regulated dc-to-dc Converter
400-V 60-W Push-Pull dc-to-dc Converter
dc-to-dc SMPS (Variable 18- to 30-V Out at 0.2A)
Minipower Inverter as a High-Voltage Low-Current Source
High-Efficiency Flyback Voltage Converter
3- to 25-V dc-to-dc Converter
Unipolar-to-Dual Supply Converter

Efficient Supply Splitter
dc-to-ac Inverter
Power Inverter (12 Vdc to 117 Vac at 60 Hz)
Power MOSFET Inverter
Medium Power Inverter
Self-Oscillating Flyback-Switching Converter
Bipolar dc-to-dc Converter
rms-to-dc Converter
Regulated dc-to-dc Converter
Positive-to-Negative Converter
Buck/Boost Converter
Isolated +15-V dc-to-dc Converter
Regulated dc-to-dc Converter
Step-Up/Step-Down dc-to-dc Converter

DC/DC REGULATING CONVERTER

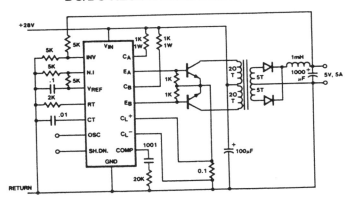

SIGNETICS

Fig. 14-1

Push-pull outputs are used in this transformer-coupled dc/dc regulating converter. Notice that the oscillator must be set at twice the desired output frequency as the SG1524's internal flip-flop divides the frequency by 2 as it switches the PWM signal from one output to the other. Current limiting is done here in the primary so that the pulse width will be reduced should transformer saturation occur.

FLYBACK CONVERTER

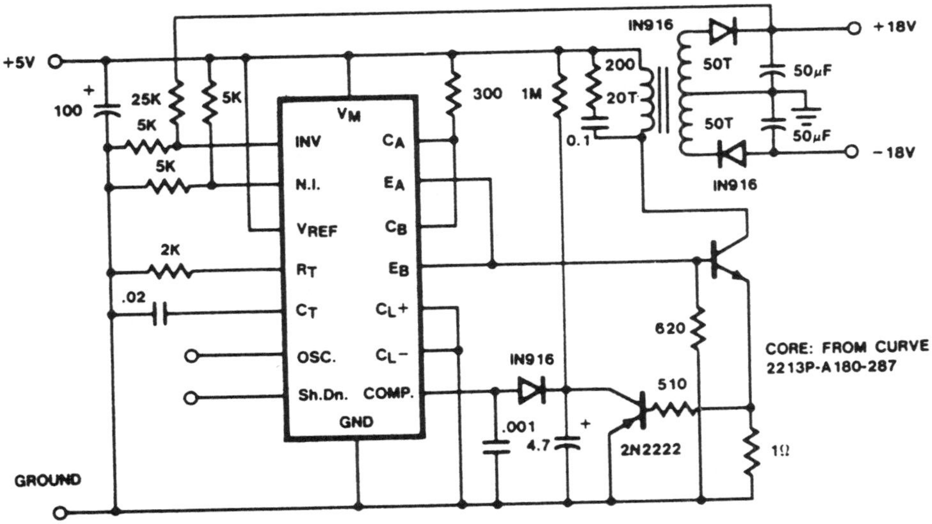

SIGNETICS

Fig. 14-2

A low-current flyback converter is used here to generate ±15 V at 20 mA from a +5-V regulated line. The reference generator in the SG1524 is unused with the input voltage providing the reference. Current limiting in a flyback converter is difficult and is accomplished here by sensing current in the primary line and resetting a soft-start circuit.

1.5-W OFFLINE CONVERTER

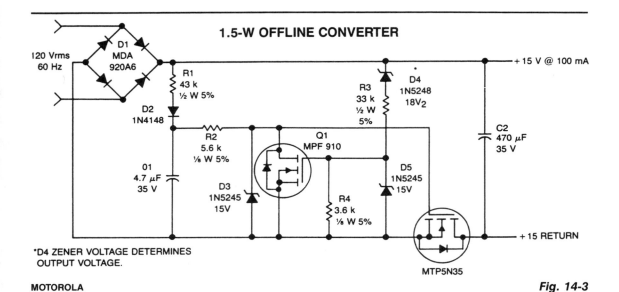

MOTOROLA

*D4 ZENER VOLTAGE DETERMINES OUTPUT VOLTAGE.

Fig. 14-3

This nonisolated unregulated minimum-component converter fills the void between low-power zener regulation and the higher power use of a 60-Hz input transformer. It is intended for use wherever a nonisolated supply can be used safely.

The circuit operates by conducting only during the low-voltage portion of the rectified sine wave. R1 and D2 charge C1 to approximately 20 V, which is maintained by Q1. This voltage is applied to the gate of Q2, turning it on. When the rectified output voltage exceeds the zener voltage of D4, Q1 turns on, shunting the gate of Q2 to ground, turning it off.

DUAL-OUTPUT ±12 or ±15 V DC/DC CONVERTER

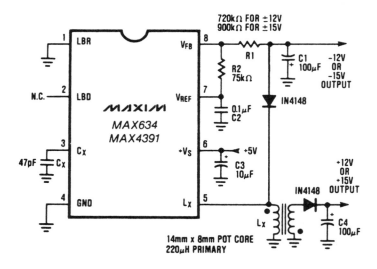

MAXIM

Fig. 14-4

DUAL-OUTPUT ±12 or ±15 V DC/DC CONVERTER *(Cont.)*

The buck-boost configuration of the MAX634 is well suited for dual output dc/dc converters. Only a second winding on the inductor is needed. Typically, this second winding is bifilar—primary and secondary are wound simultaneously using two wires in parallel. The inductor core is usually a toroid or a pot core. The negative output voltage is fully regulated by the MAX634. The positive voltage is semiregulated, and will vary slightly with load changes on either the positive or negative outputs.

12-V CONVERTER

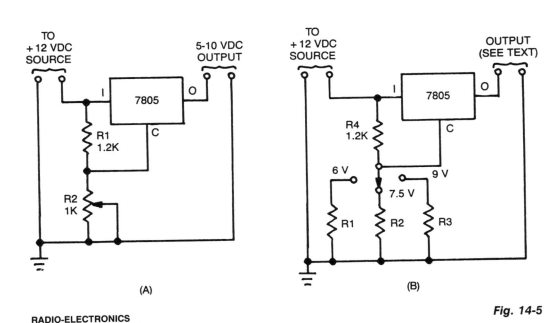

(A)

(B)

Fig. 14-5

Many devices operate from a car's 12-V electrical system. Some require 12 V; others require some lesser voltage. An automobile battery's output can vary from 12 to 13.8 V under normal circumstances. The load requirements of the device might vary. This circuit maintains a constant voltage, regardless of how those factors change. Simple circuit, A, uses a 7805 voltage regulator. In addition to a constant output, this IC provides overload and short-circuit protection. That unit is a 5-V, 1-A regulator, but when placed in circuit B, it can provide other voltages as well. When the arm of potentiometer R1 is moved toward ground, the output varies from 5 to about 10 V.

REGULATED 6-V to 15-V FLYBACK CONVERTER

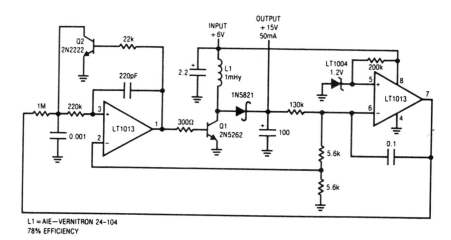

L1 = AIE—VERNITRON 24-104
78% EFFICIENCY

LINEAR TECHNOLOGY

Fig. 14-6

This converter delivers up to 50 mA from a 6-V battery with 78% efficiency. This flyback converter functions by feedback-controlling the frequency of inductive flyback events. The inductor's output, rectified and filtered to dc, biases the feedback loop to establish a stable output. If the converter's output is below the loop setpoint, A2's inputs unbalance and current is fed through the 1-MΩ resistor at A1. This ramps the 1000-pF value positive. When this ramp exceeds the 0.5-V potential at A1's positive input, the amplifier switches high. Q2 turns on, discharging the capacitor to ground. Simultaneously, regenerative feedback through the 200-pF value causes a positive-going pulse at A1's positive input, sustaining A1's positive output. Q1 comes on, allowing inductor L1 current to flow. When A1's feedback pulse decays, its output becomes low, turning off Q1. Q1's collector is pulled high by the inductor's flyback and the energy is stored in the 100-μF capacitor. The capacitor's voltage, which is the circuit output, is sampled by A2 to close to loop around A1/Q1. This loop forces A1 to oscillate at whatever frequency is required to maintain the 15-V output.

In-phase transformer windings for the drain and gate of TMOS power FET Q1 cause the circuit to oscillate. Oscillation starts when the feedback coupling capacitor, C1, is charged from the supply line via a large resistance; R2 and R3 limit the collector current to Q2. During *pump-up*, the one time is terminated by Q2, which senses the ramped source current of Q1. C1 is charged on alternate half-cycles by Q2 and forward-biased by zener D2.

When the regulated level is reached, forward bias is applied to Q2, terminating the on time earlier at a lower peak current. When this occurs, the frequency increases in inverse proportion to current, but the energy per cycle decreases in proportion to current squared. Therefore, the total power coupled through the transformer to the secondary is decreased.

REGULATED DC-TO-DC CONVERTER

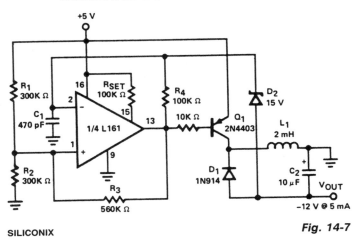

SILICONIX

Fig. 14-7

This low power dc-to-dc converter is obtained by adding a flyback circuit to a square-wave oscillator. The operating frequency is 20 kHz to minimize the size of L1 and C2. Regulation is achieved by zener diode D2. Maximum current available before the converter drops out of regulation is 5.5 mA.

400-V 60-W PUSH-PULL DC-TO-DC CONVERTER

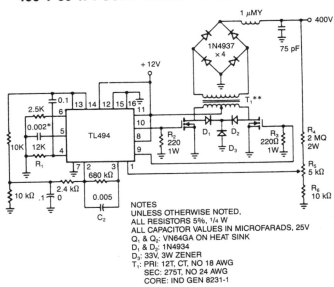

SILICONIX

NOTES
UNLESS OTHERWISE NOTED,
ALL RESISTORS 5%, 1/4 W
ALL CAPACITOR VALUES IN MICROFARADS, 25V
Q_1 & Q_2: VN64GA ON HEAT SINK
D_1 & D_2: 1N4934
D_3: 33V, 3W ZENER
T_1: PRI: 12T, CT, NO 18 AWG
        SEC: 275T, NO 24 AWG
        CORE: IND GEN 8231-1

Fig. 14-8

The TL494 switching regulator governs the operating frequency and regulates output voltage. The switching frequency is approximately 100 kHz for the values shown. Output regulation is typically 1.25% from no-load to full 60 W.

DC-TO-DC SMPS (VARIABLE 18- to 30-V OUT AT 0.2 A)

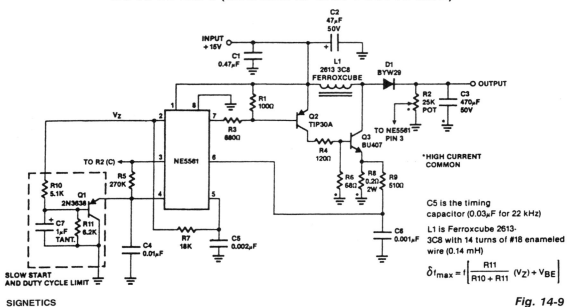

SIGNETICS

Fig. 14-9

MINIPOWER INVERTER AS A HIGH-VOLTAGE LOW-CURRENT SOURCE

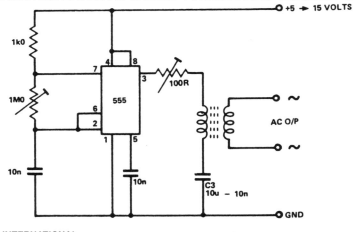

ELECTRONICS TODAY INTERNATIONAL

Fig. 14-10

The circuit is capable of providing power for portable Geiger counters, dosimeter chargers, high-resistance meters, etc. The 555 timer IC is used in its multivibrator mode, the frequency adjusted to optimize the transformer characteristics. When the output of the IC is high, current flows through the limiting resistor, the primary coil to charge C3. When the output is low, the current is reversed. With a suitable choice of frequency and C3, a good symmetric output is sustained.

HIGH-EFFICIENCY FLYBACK VOLTAGE CONVERTER

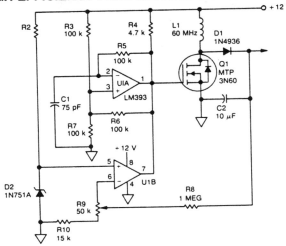

Fig. 14-11

MOTOROLA

U1 is a dual voltage comparator with open collector outputs. The A side is an oscillator operating at 100 kHz, and the B side is part of the regulation circuit that compares a fraction of the output voltage to a reference generated by zener diode D2.

The output of U1A is applied directly to the gate of Q1. During the positive half-cycle of the Q1 gate voltage, energy is stored in L1; in the negative half, the energy is discharged into C2. A portion of the output voltage is fed back to U1B to provide regulation. The output voltage is adjustable by changing feedback potentiometer R9.

Using the component values shown will produce a nominal 300-V output from a 12-V source. However, the circuit maximum output voltage is limited by R10; a lower value for R10 yields a higher output voltage. The output voltage is also limited by the breakdown of values Q1, L1, D1, and C2.

3-TO 25-V DC-TO-DC CONVERTER

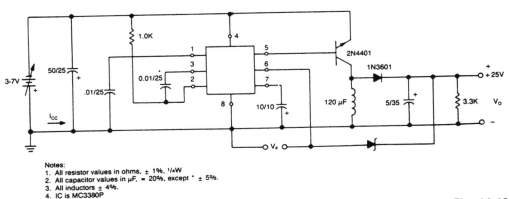

Notes:
1. All resistor values in ohms, ± 1%, ¼W.
2. All capacitor values in μF, ≈ 20%, except * ± 5%.
3. All inductors ± 4%.
4. IC is MC3380P

MOTOROLA

Fig. 14-12

143

UNIPOLAR-TO-DUAL SUPPLY CONVERTER

The outputs in this circuit are independently variable and can be loaded unsymmetrically. The output voltage remains constant, irrespective of load and changes. By varying potentiometers R2 or R6, the output voltages can be conveniently set. Outputs can be varied between 8 and 17 V so that the standard ±9-, ±12-, and ±15-V settings can be made. This converter is designed for a maximum load current of 1 A and the output impedance of both supplies of 0.35 Ω. This circuit is not protected against short circuits, but uses the protection provided by the dc input source. This circuit is ideal for biasing operation amplifier circuits.

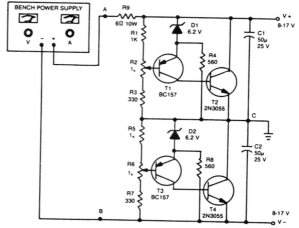

ELECTRONIC ENGINEERING

Fig. 14-13

EFFICIENT SUPPLY SPLITTER

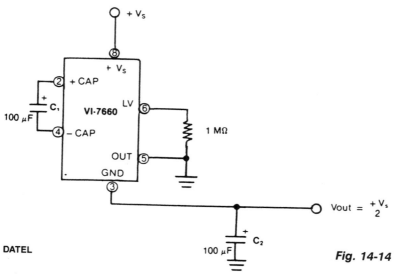

DATEL

Fig. 14-14

In this application, the VI-7660 is connected as a voltage splitter. Notice that the *normal* output pin is connected to ground and the *normal* ground pin is used as the output. The switches that allow the charge pumping are bidirectional; therefore, charge transfer can be performed in reverse. The 1-MΩ resistor is used to avoid start-up problems by forcing the internal regulator on. An application for this circuit would be driving low-voltage ±7.5-Vdc circuits from ±15-Vdc supplies, or driving low-voltage logic from 9- to 12-V batteries.

DC-TO-AC INVERTER

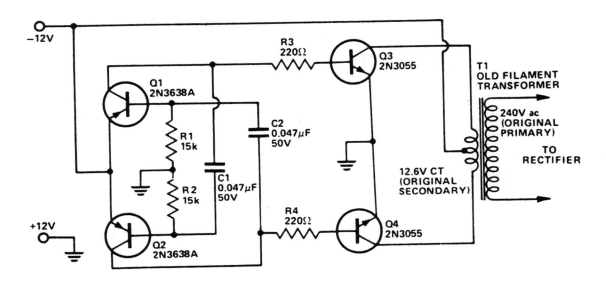

Fig. 14-15

This inverter uses no special components, such as the torodial transformer used in many inverters. Cost is kept low with the use of cheap, readily available components. Essentially, it is a power amplifier that is driven by an astable multivibrator. The frequency is around 1200 Hz, which most 50-/60-Hz power transformers handle well without too much loss. Increasing the value of capacitors C1 and C2 will lower the frequency if any trouble is experienced. However, the rectifier filtering capacitors required are considerably smaller at the higher operating frequency. The two 2N3055 transistors should be mounted on an adequately sized heatsink. The transformer should be rated according to the amount of output power required, allowing for conversion efficiency of approximately 60%.

POWER INVERTER (12 Vdc TO 117 Vac AT 60 Hz)

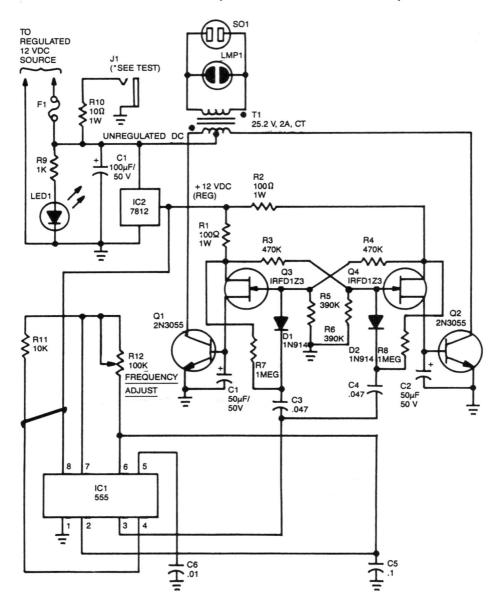

Fig. 14-16

POWER INVERTER (12 Vdc TO 117 Vac AT 60 Hz) *(Cont.)*

Capacitor C5 and potentiometer R12 determine the frequency of the output signal at pin 3 of IC1, the 555 oscillator. The output signal is differentiated by C3 and C4 before it is input to the base of power transistors Q1 and Q2 via diodes D1 and D2, respectively. The signal from IC1 is adjusted to 120 Hz because the flip-flop formed by transistors Q3 and Q4 divides the frequency by 2.

When Q3 is on, the base of Q1 is connected via R1 to the regulated 12-V supply. Then, when the flip-flop changes states, Q4 is turned on and the base of Q2 is connected to the 12-V supply through R2. The 100-mA base current allows Q1 and Q2 to alternately conduct through their respective halves to the transformer's secondary winding.

To eliminate switching transients caused by the rapid switching of Q3 and Q4, capacitors C1 and C2 filter the inputs to the base of Q1 and Q2 respectively. Power for the unit comes from an automobile's 12-V system or from a storage battery. The power is regulated by IC2, a 7812 regulator. LED1, connected across the 12-V input, can be used to indicate whether power is being fed to the circuit. The neon pilot lamp, LMP1, shows a presence or absence of output power.

POWER MOSFET INVERTER

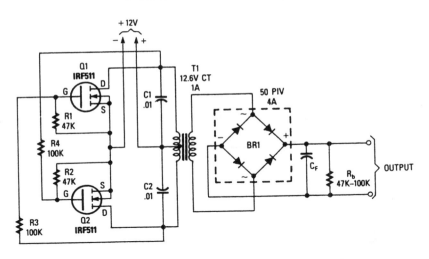

POPULAR ELECTRONICS

Fig. 14-17

This inverter can deliver high-voltage ac or dc, with a rectifier and filter, up to several hundred volts. The secondary and primary of T1—a 12.6- to 440-V power transformer, respectively—are reversed (e.g., the primary becomes the secondary and the secondary becomes the primary). Transistors Q1 and Q2 can be any power FET. Be sure to heatsink Q1 and Q2. Capacitors C1 and C2 are used as spike suppressors.

MEDIUM POWER INVERTER

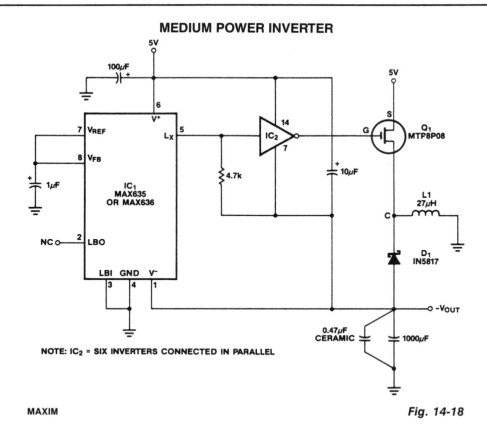

NOTE: IC₂ = SIX INVERTERS CONNECTED IN PARALLEL

MAXIM   *Fig. 14-18*

In this circuit, a CMOS inverter, such as the CD4069, is used to convert the open drain L_x output to a signal that is suitable for driving the gate of an external P MOSFET. The MTP8P03 has a gate threshold voltage of 2.0 to 4.5 V, so it will have a relatively high resistance if driven with only 5 V of gate drive. To increase the gate drive voltage, and thereby increase efficiency and power handling capability, the negative supply pin of the CMOS inverter is connected to the negative output, rather than to ground. Once the circuit is started, the P MOSFET gate drive swings from +5 V to $-V_{OUT}$. At start-up, the $-V_{OUT}$ is one Schottky diode drop above ground and the gate drive to the power MOSFET is slightly less than 5 V. The output should be only lightly loaded to ensure start-up, because the output power capability of the circuit is very low until $-V_{OUT}$ is a couple of volts.

This circuit generates complementary output signals that range from 50 to 240 Hz. Digital-timing control ensures a separation of 10 to 15° between the fall time of one output and the rise time of the complementary output.

The digital portion of inverter U1 to U4 controls the drive to Q1 and Q2, both MTE60N20 TMOS devices. These devices are turned on alternately with 11.25° separation between complementary outputs. A +12-V supply for CMOS gates U1 to U4 is developed by T1, D3, D4, C7, and U6. The power supply for the TMOS frequency generator is derived from the diode bridge, U5, and capacitor C7; it is applied to the center tap of T2.

SELF-OSCILLATING FLYBACK-SWITCHING CONVERTER

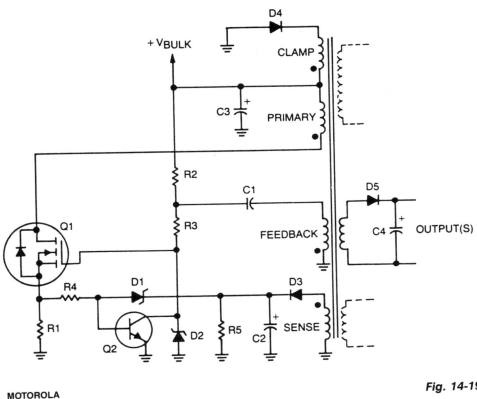

MOTOROLA

Fig. 14-19

Regulation is provided by taking the rectified output of the sense winding and applying it as a bias to the base of Q2 via zener D1. The collector of Q2 then removes drive from the gate of Q1. Therefore, if the output voltage should increase, Q2 removes the drive to Q1 earlier, shortening the on time, and the output voltage will remain the same. dc outputs are obtained by merely rectifying and filtering secondary windings, as done by D5 and C4.

BIPOLAR DC-TO-DC CONVERTER

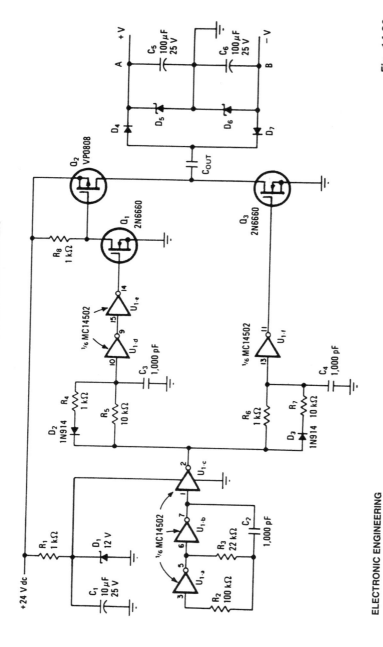

Fig. 14-20

ELECTRONIC ENGINEERING

Inverters U1a and U1b form a 20-kHz oscillator whose square-wave output (further shaped by D2, R4, and R5, and by D3, R6, and R7) drives power FETs Q2 and Q3. The p-channel and n-channel FETs conduct alternately, in a push-pull configuration. When Q2 conducts, the positive charge on C_{out} forces diode D4 to conduct as well, which provides a positive voltage, determined by zener diode D5, at terminal A. Similarly, when Q3, in its turn conducts, the negative charge on C_{out} forces D7 to do so as well. A negative voltage, therefore, develops at terminal B, whose level is set by D6.

RMS-TO-DC CONVERTER

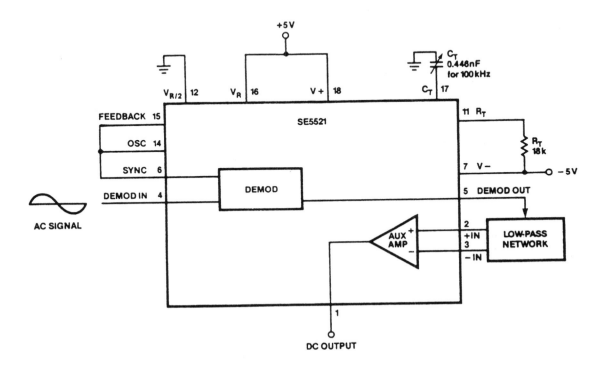

NOTE:
1. The DC output at Pin 1 varies linearly with the RMS input at Pin 4.
2. C_T is tweaked until the sync signal is in phase with the AC signal.

SIGNETICS

Fig. 14-21

An ac voltmeter can be easily constructed. The simplicity of the circuit and the low component count make it particularly attractive. The demodulator output is a full-wave rectified signal from the ac input at Pin 4. The dc component on the rectified signal at Pin 5 varies linearity with the rms input at Pin 4 and thus provides an accurate rms-to-dc conversion at the output of the filter (Pin 1). C_T is a variable capacitor that is tweaked until the oscillator signal to the sync input of the demodulator is in phase with the ac signal at Pin 4.

REGULATED DC-TO-DC CONVERTER

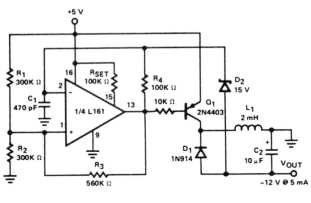

SILICONIX

Fig. 14-22

POSITIVE-TO-NEGATIVE CONVERTER

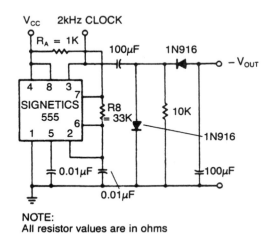

NOTE:
All resistor values are in ohms

(a) POSITIVE-TO-NEGATIVE CONVERTER

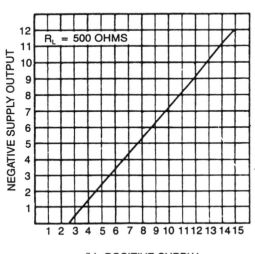

(b) POSITIVE SUPPLY

SIGNETICS

Fig. 14-23

POSITIVE-TO-NEGATIVE CONVERTER *(Cont.)*

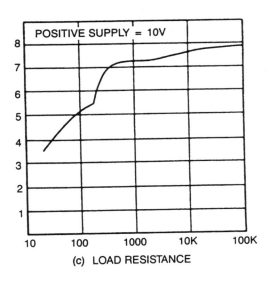

POSITIVE SUPPLY = 10V

(c) LOAD RESISTANCE

The transformerless dc/dc converter derives a negative supply voltage from a positive. As a bonus, the circuit also generates a clock signal. The negative output voltage tracks the dc-input voltage linearity (a), but its magnitude is about 3 V lower. Application of a 500-Ω load, (b), causes 10% change from the no-load value.

BUCK/BOOST CONVERTER

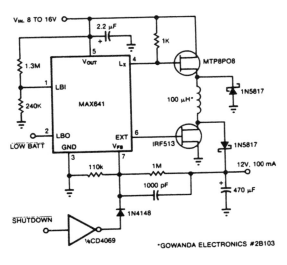

This converter can accommodate wide input-voltage swings, such as the 8- to 15-V swing typical of a 12-V sealed lead/acid battery. The low-battery output indicates when the input voltage drops below 8 V. Pulling shutdown turns off the circuit.

MAXIM

Fig. 14-24

ISOLATED +15-V DC-TO-DC CONVERTER

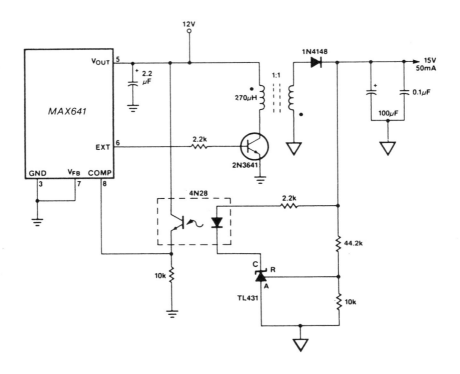

MAXIM

Fig. 14-25

In this circuit, a TL431 shunt regulator is used to sense the output voltage. The TL431 drives the LED of a 4N28 optocoupler which provides feedback to the MAX641 while maintaining isolation between the input, +12 V, and the output, +15 V. In this circuit, the +15-V output is fully regulated, with respect to both line and load changes.

REGULATED DC-TO-DC CONVERTER

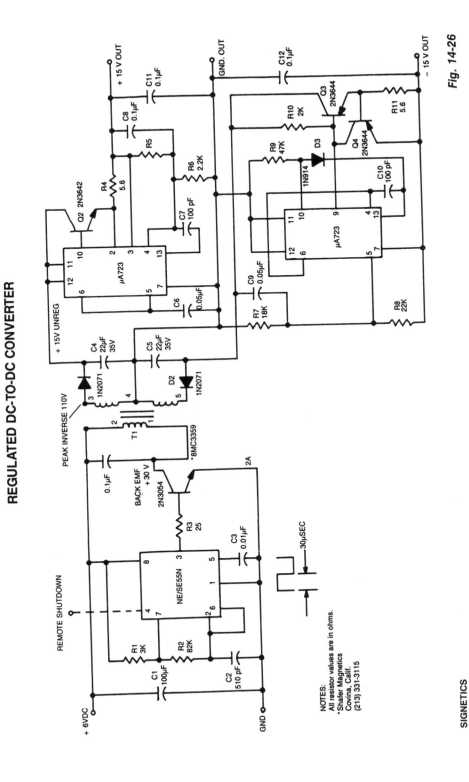

SIGNETICS

Fig. 14-26

The regulated dc/dc converter produces 15-Vdc outputs from a +5 Vdc input. Line and load regulation is 0.1%.

STEP-UP/STEP-DOWN DC-TO-DC CONVERTER

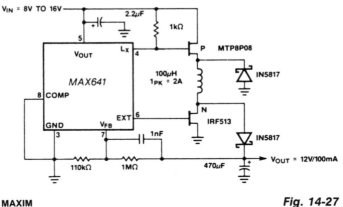

Fig. 14-27

Positive output step-up and step-down dc/dc converters have a common limitation in that neither can handle input voltages that are both greater than or less than the output. For example, when converting a 12-V sealed lead-acid battery to a regulated +12-V output, the battery voltage might vary from a high of 15 V down to 10 V.

By using a MAX641 to drive separate p- and n-channel MOSFETs, both ends of the inductor are switched to allow noninverting buck/boost operation. A second advantage of the circuit over most boost-only designs is that the output goes to 0 V when shutdown is activated. Inefficiency is a drawback because two MOSFETs and two diodes increase the losses in the charge and discharge path of the inductor. The circuit delivers +12 V at 100 mA at 70 percent efficiency with an 8-V input.

15

Temperature-to-Frequency Converters

The sources of the following circuits are contained in the Sources section, which begins on page 175. The figure number in the box of each circuit correlates to the source entry in the Sources section.

LINEAR TEMPERATURE-TO-FREQUENCY CONVERTER

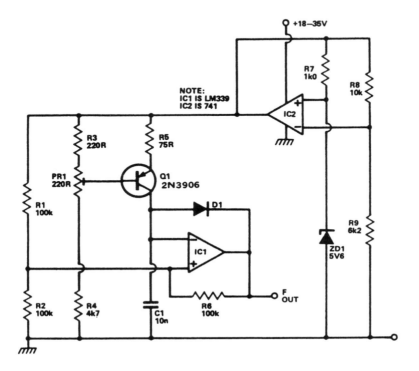

ELECTRONICS TODAY INTERNATIONAL

Fig. 15-1

This circuit provides a linear increase of frequency of 10 Hz/°C over 0 to 100°C and can thus be used with logic systems, including microprocessors. Temperature probes Q1 V_{be} changes 2.2 mV/°C. This transistor is incorporated in a constant current-source circuit. Thus, a current proportional to temperature will be available to charge C1. The circuit is powered via the temperature-stable reference voltage that is supplied by the 741. Comparator IC1 is used as a Schmitt trigger whose output is used to discharge C1 via D1. To calibrate the circuit, Q1 is immersed in boiling distilled water and PR1 is adjusted to give 1-kHz output. The prototype was found to be accurate to within 0.2 °C.

TEMPERATURE-TO-FREQUENCY CONVERTER I

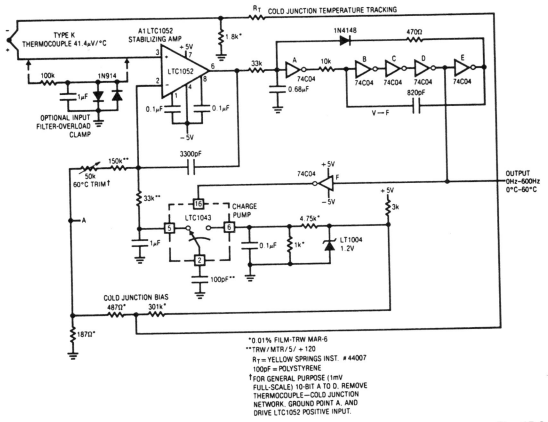

LINEAR TECHNOLOGY

Fig. 15-2

A1's positive input is biased by the thermocouple. A1's output drives a crude V/F converter, comprised of the 74C04 inverters and associated components. Each V/F output pulse causes a fixed quantity of charge to be dispensed into the 1-μF capacitor from the 100-pF capacitor via the LT1043 switch. The larger capacitor integrates the packets of charge, producing a dc voltage at A1's negative input. A1's output forces the V/F converter to run at whatever frequency is required to balance the amplifier's inputs. This feedback action eliminates drift and nonlinearities in the V/F converter as an error item and the output frequency is solely a function of the dc conditions at A1's inputs. The 3300-pF capacitor forms a dominant-response pole at A1, stabilizing the loop.

TEMPERATURE-TO-FREQUENCY CONVERTER II

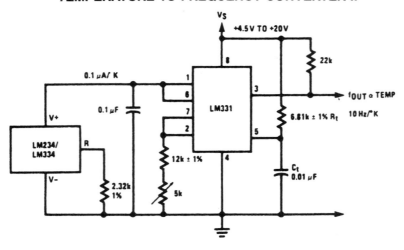

NATIONAL SEMICONDUCTOR

Fig. 15-3

TEMPERATURE-TO-FREQUENCY CONVERTER III

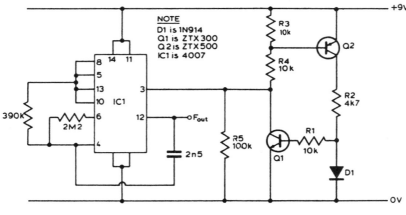

ELECTRONICS TODAY INTERNATIONAL

Fig. 15-4

This circuit exploits the forward voltage of a silicon diode that varies with temperature in a reasonably linear way when fed from a constant-current source. Diode D1 and resistor R2 form a potential divider that is fed from the constant-current source. As the temperature rises, the forward voltage of D1 falls, and turns Q1 off. The output voltage from Q1 will thus rise, and this is used as the control voltage for the CMOS VCO. With the values shown, the device gave an increase of just under 3 Hz/°C (between 0°C and 60°C) giving a frequency of 470 Hz at 0°C.

16

Voltage-to-Current Converters

The sources of the following circuits are contained in the Sources section, which begins on page 175. The figure number in the box of each circuit correlates to the source entry in the Sources section.

VOLTAGE-TO-CURRENT CONVERTER I

The current out is $I_{OUT} \cong V_{IN}/R$. For negative currents, a pnp can be used and, for better accuracy, a Darlington pair can be substituted for the transistor. With careful design, this circuit can be used to control currents to many amps. Unity-gain compensation is necessary.

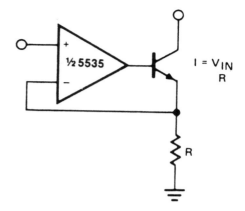

SIGNETICS

Fig. 16-1

VOLTAGE-TO-CURRENT CONVERTER II

A simple voltage-to-current converter is shown in the figure. The current out is I_{OUT} or V_{IN}/R. For negative currents, a pnp can be used and, for better accuracy, a Darlington pair can be substituted for the transistor. With careful design, this circuit can be used to control currents of many amps. Unity-gain compensation is necessary.

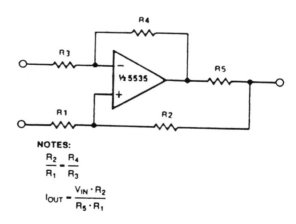

NOTES:

$$\frac{R_2}{R_1} = \frac{R_4}{R_3}$$

$$I_{OUT} = \frac{V_{IN} \cdot R_2}{R_5 \cdot R_1}$$

SIGNETICS

Fig. 16-2

162

17

Voltage-to-Frequency Converters

The sources of the following circuits are contained in the Sources section, which begins on page 175. The figure number in the box of each circuit correlates to the source entry in the Sources section.

LOW-COST VOLTAGE-TO-FREQUENCY CONVERTER

The 741 op-amp integrator signal is fed into the Schmitt-trigger input of an inverter. When the signal reaches the magnitude of the positive-going threshold voltage, the output of the inverter is switched to 0. The inverter output controls the FET switch directly. For a gate voltage of 0, the FET channel turns on to low resistance and the capacitor is discharged. The discharge current depends on the resistance of the FET. When the capacitor C1 is discharged to the negative-going threshold voltage level of the inverter, the inverter output is switched to ±12 V. This switch causes the FET channel to be switched off, and the discharging process is switched into a charging process again. Using the components shown, an output frequency of about 10 kHz with 0.1% linearity can be obtained.

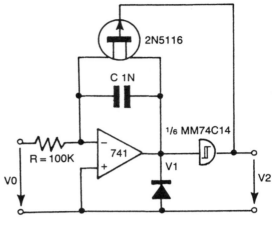

ELECTRONIC ENGINEERING

Fig. 17-1

WIDE-RANGE VOLTAGE-TO-FREQUENCY CONVERTER

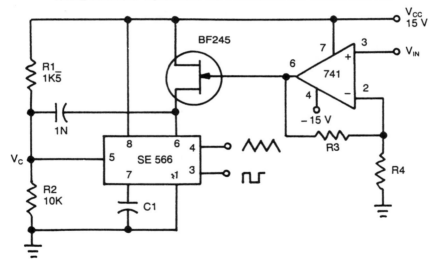

ELECTRONIC ENGINEERING

Fig. 17-2

This circuit is based upon the change of frequency of the function generator with the input voltage V_{IN}. Generally, the frequency depends upon the capacitance and resistor connected to pin 6. This resistor is replaced by the FET. The frequency range is adjustable by changing the input voltage, V_{IN}; the converter will give a range of 10 Hz to 1 MHz.

ULTRAPRECISION V/F CONVERTER

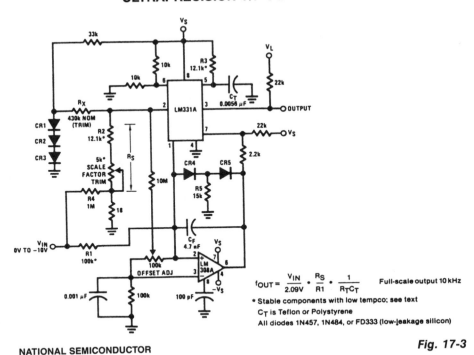

$$f_{OUT} = \frac{V_{IN}}{2.09V} \cdot \frac{R_S}{R1} \cdot \frac{1}{R_T C_T} \qquad \text{Full-scale output 10 kHz}$$

* Stable components with low tempco; see text
C_T is Teflon or Polystyrene
All diodes 1N457, 1N484, or FD333 (low-leakage silicon)

NATIONAL SEMICONDUCTOR

Fig. 17-3

The circuit is capable of better than 0.02% error and 0.003% nonlinearity for a ±20°C range about room temperature.

V/F CONVERTER (NEGATIVE INPUT VOLTAGE)

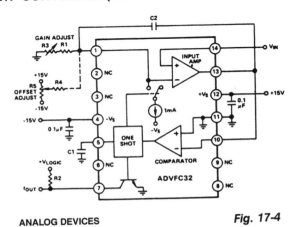

ANALOG DEVICES

Fig. 17-4

PRESERVED-INPUT VOLTAGE-TO-FREQUENCY CONVERTER

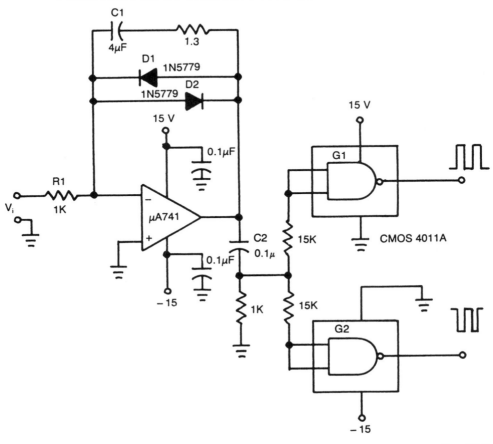

Fig. 17-5

The input voltage, V_1, causes C1 to charge and produce a ramp voltage at the output of the 741 op amp. Diodes D1 and D2 are four-layer devices. When the voltage across C1 reaches the breakover voltage of either diode, the diode conducts to discharge C1 rapidly and the op-amp output goes abruptly to 0. This rapid discharge action applies a narrow pulse to G1 and G2. Positive discharge pulses produced by a positive V_1 are coupled to the output only through G1; negative pulses are coupled only through G2.

Because of the forward breakover current of diodes D1 and D2, the circuit won't operate below a minimum input voltage. An increase of R_1 increases this minimum voltage and reduces the circuit's dynamic range. The minimum input voltage with R1 at 1 kΩ is in the range of 10 to 50 mV. This input dead zone, when input signal V_1 is near 0, is desirable in applications that require a signal to exceed a certain level before an output is generated.

PRESERVED-INPUT VOLTAGE-TO-FREQUENCY CONVERTER *(Cont.)*

This circuit can accept positive or negative or differential control voltages. The output frequency is 0 when the control voltage is 0. The 741 op amp forms a current source controlled by the voltage E_C to charge the timing capacitor C1 linearly. NE555 is connected in the astable mode, so that the capacitor charges and discharges between $1/3\ V_{CC}$ and $2/3\ V_{CC}$. The offset is adjusted by the 10-k potentiometer so that the frequency is 0 when the input is 0. For the component values shown: $f \approx 4.2\text{-}E_C$ kHz. If two dc voltages are applied to the ends of R1 and R4, the output frequency will be proportional to the difference between the two voltages.

1-Hz TO 1.25-MHz VOLTAGE-TO-FREQUENCY CONVERTER

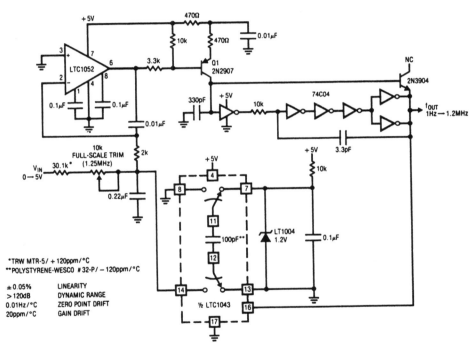

LINEAR TECHNOLOGY

Fig. 17-6

This stablized voltage-to-frequency converter features 1-Hz to 1.25-MHz operation, 0.05% linearity, and a temperature coefficient of typically 20 ppm/°C. This circuit runs from a single 5-V supply. The converter uses a charge feedback scheme to allow the LTC1052 to close a loop around the entire circuit, instead of simply controlling the offset. This approach enhances linearity and stability, but introduces the loop's settling time into the overall voltage-to-frequency step-response characteristic.

10-Hz TO 10-kHz V/F CONVERTER

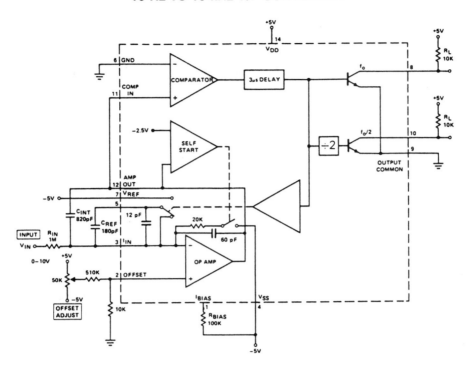

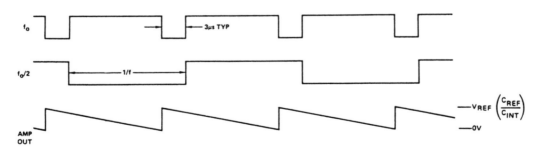

Notes:
1. To adjust f_{min}, set V_{IN} = 10 mV and adjust the 50 k offset for 10 Hz out.
2. To adjust f_{max}, set V_{IN} = 10 V and adjust R_{IN} or V_{REF} for 10 kHz out.
3. To increase f_{OUT} MAX to 100 kHz change C_{REF} to 27 pF and C_{INT} to 75 pF.
4. For high performance applications use high stability components for R_{IN}, C_{REF}, V_{REF} (metal film resistors and glass film capacitors). Also separate the output ground (Pin 9) from the input ground (Pin 6).

ACCURATE VOLTAGE-TO-FREQUENCY CONVERTER

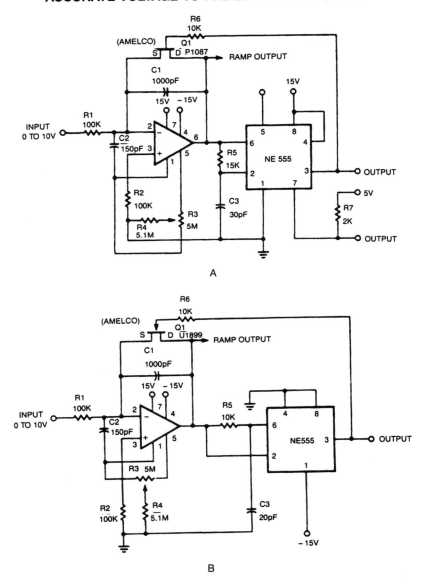

A

B

SIGNETICS

Fig. 17-8

This linear voltage-to-frequency converter, A, achieves good linearity over 0 to −10 V. Its mirror image, B, provides the same linearity over 0 to +10 V, but it is not DTL/TTL compatible.

VOLTAGE-TO-FREQUENCY CONVERTER I

The D169 serves as a level detector and provides complementary outputs. The op amp is used to integrate the input signal V_{IN} with a time constant of R_1C_1. The input (must be negative) causes a positive ramp at the output of the integrator, which is summed with a negative zener voltage. When the ramp is positive enough, D169 outputs change state and out 2 flips from negative to positive. The output-pulse repetition rate f_o, is directly proportioned to the negative input voltage, V_{IN}.

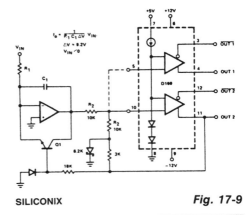

SILICONIX

Fig. 17-9

VOLTAGE-TO-FREQUENCY CONVERTER II

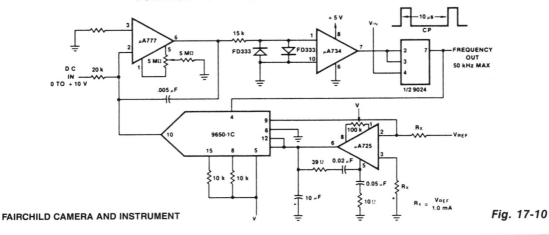

FAIRCHILD CAMERA AND INSTRUMENT

Fig. 17-10

V/F CONVERTER (POSITIVE INPUT VOLTAGE)

ANALOG DEVICES

Fig. 17-11

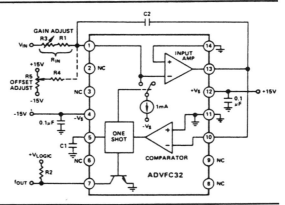

PRECISION VOLTAGE-TO-FREQUENCY CONVERTER

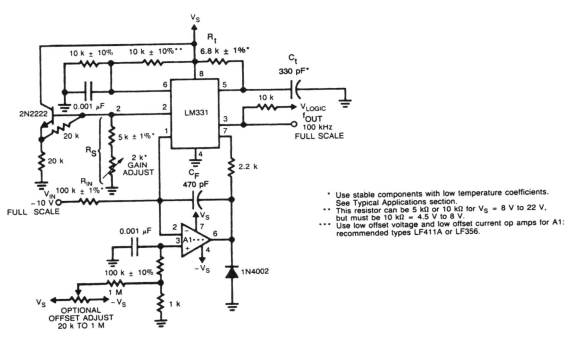

* Use stable components with low temperature coefficients.
  See Typical Applications section.
** This resistor can be 5 kΩ or 10 kΩ for V_S = 8 V to 22 V,
   but must be 10 kΩ = 4.5 V to 8 V.
*** Use low offset voltage and low offset current op amps for A1:
   recommended types LF411A or LF356.

NATIONAL SEMICONDUCTOR

Fig. 17-12

In this circuit, integration is performed by using a conventional operational amplifier and feedback capacitor, C_F. When the integrator's output crosses the nominal threshold level at pin 6 of the LM131, the timing cycle is initiated. The average current fed into the op amp's summing point (pin 2) is $i \times (1.1 R_t C_t) \times f$ which is perfectly balanced with $-V_{IN}/R_{IN}$. In this circuit, the voltage offset of the LM131 input comparator does not affect the offset or accuracy of the V/F converter as it does in the stand-alone V/F converter, nor does the LM131 bias current or offset current. Instead, the offset voltage and offset current of the operational amplifier are the only limits on how small the signal can be accurately converted.

1-Hz TO 30-MHz VOLTAGE-TO-FREQUENCY CONVERTER

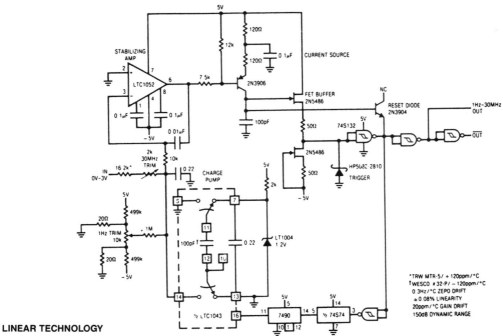

LINEAR TECHNOLOGY

Fig. 17-13

Circuit has a 1-Hz to 30-MHz output, 150-dB dynamic range, for a 0- to 5-V input. It maintains 0.08% linearity over its entire 7$\frac{1}{3}$-decade range with a full-scale drift of about 20 ppm/°C. To get the additional bandwidth, the fast JFET buffer drives the Schottky TTL Schmitt trigger. The Schottky diode prevents the Schmitt trigger from ever seeing a negative voltage at its input. The Schmitt's input voltage hysteresis provides the limits which the oscillator runs between. The 30-MHz, full-scale output is much faster than the LTC1043 can accept, so the digital divider stages are used to reduce the feedback frequency signal by a factor of 20. The remaining Schmitt sections furnish complementary outputs.

DIFFERENTIAL-INPUT VOLTAGE-TO-FREQUENCY CONVERTER

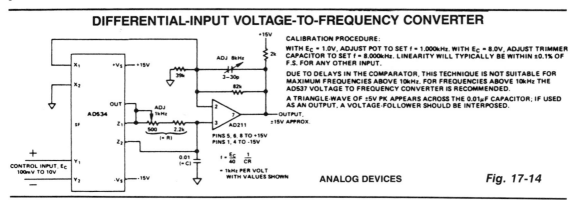

CALIBRATION PROCEDURE:

WITH E_C = 1.0V, ADJUST POT TO SET f = 1.000kHz. WITH E_C = 8.0V, ADJUST TRIMMER CAPACITOR TO SET f = 8.000kHz. LINEARITY WILL TYPICALLY BE WITHIN ±0.1% OF F.S. FOR ANY OTHER INPUT.

DUE TO DELAYS IN THE COMPARATOR, THIS TECHNIQUE IS NOT SUITABLE FOR MAXIMUM FREQUENCIES ABOVE 10kHz. FOR FREQUENCIES ABOVE 10kHz THE AD537 VOLTAGE TO FREQUENCY CONVERTER IS RECOMMENDED.

A TRIANGLE-WAVE OF ±5V PK APPEARS ACROSS THE 0.01μF CAPACITOR; IF USED AS AN OUTPUT, A VOLTAGE-FOLLOWER SHOULD BE INTERPOSED.

ANALOG DEVICES

Fig. 17-14

VOLTAGE-TO-FREQUENCY CONVERTER III

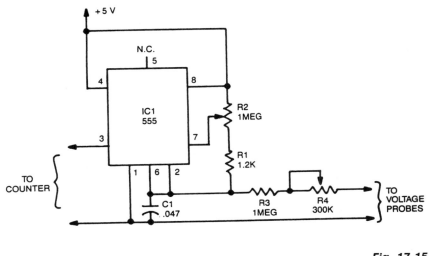

GERNSBACK

Fig. 17-15

The output frequency from IC pin 3 is determined by the voltage input to pin 6. A standard frequency counter can be used to measure voltages directly over a limited range from 0 to 5 V. In this circuit, the 555 is wired as an astable multivibrator. Resistor R2 determines the output frequency when the input to the circuit (the voltage measured by the voltage probes) is 0. R4 is a scaling resistor that adjusts the output frequency so that a change in the input voltage of 1 V will result in a change in the output frequency of 10 Hz. That will happen when the combined resistance of R3 and R4 is 1.2 MΩ. To calibrate, short the voltage probes together and adjust R2 until the reading on the frequency counter changes to 00 Hz. Then, use the voltage probes to measure an accurate 5-V source and adjust R4 until the frequency counter reads 50 Hz.

VOLTAGE-TO-FREQUENCY CONVERTER IV

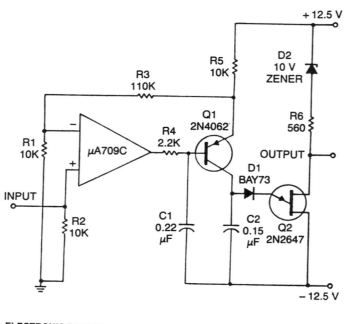

Fig. 17-16

This circuit consists of a UJT oscillator in which the timing charge capacitor C2 is linearly dependent on the input signal voltage. The charging current is set by the voltage across resistor R5, which is accurately controlled by the amplifier.

Sources

Chapter 1

Fig. 1-1. Linear Technology Corp., Linear Applications Handbook, 1987, p. AN15-2.

Fig. 1-2. Courtesy of Fairchild Camera & Instrument Corporation. Linear Databook, 1982, p. 7-8.

Fig. 1-3. National Semiconductor Corp., 1984 Linear Supplement Databook, p. S5-126.

Fig. 1-4. Intersil, Data Book, 5/83, p. 4-83.

Fig. 1-5. Ferranti, Technical Handbook Vol. 10, Data Converters, 1983, p. 7-10.

Fig. 1-6. Precision Monolithics Inc., 1981 Full Line Catalog, p. 16-12.

Fig. 1-7. Signetics, 1987 Linear Data Manual, Vol. 2: Industrial, 11/86, p. 5-215.

Fig. 1-8. Signetics, 1987 Linear Data Manual, Vol. 2: Industrial, 12/86, p. 4-67.

Fig. 1-9. Harris, Analog Product Data Book, 1988, p. 10-48.

Fig. 1-10. Linear Technology Corp., Linear Applications Handbook, 1987, p. AN15-2.

Fig. 1-11. Reprinted with permission of Analog Devices, Inc., Data Acquisition Databook, 1982, p. 10-241.

Fig. 1-12. Precision Monolithics Inc., 1981 Full Line Catalog, p. 8-13.

Fig. 1-13. Signetics, 1987 Linear Data Manual Vol 2: Industrial, 2/87, p. 5-311.

Fig. 1-14. Reprinted with the permission of National Semiconductor Corp., National Semiconductor CMOS Databook. 1981. p. 3-63.

Fig. 1-15. Reprinted with permission of Analog Devices, Inc., Data Acquisition Databook, 1982, p. 10-240.

Fig. 1-16. Teledyne Semiconductor, Data and Design Manual, 1981, p. 7-39.

Fig. 1-17. Reprinted with permission of Analog Devices, Inc., Data Acquisition Databook, 1982, p. 10-50.

Fig. 1-18. Courtesy of Fairchild Camera and Instrument Corp., Linear Databook, 1982, p. 5-32.

Fig. 1-19. Precision Monolithics Inc., 1981 Full Line Catalog, p. 8-13.

Fig. 1-20. Siliconix, Integrated Circuits Data Book, 1988, p. 6-148.

Fig. 1-21. Intersil, Component Data Catalog, 1987, p. 4-43.

Fig. 1-22. Linear Technology Corp., Application Note 9, p. 16.

Fig. 1-23. Maxim, Seminar Applications Book, 1988/89, p. 38.

Fig. 1-24. Linear Technology Corp., Linear Databook, 1986, p. 5-17.

Fig. 1-25. GE/RCA, BiMOS Operational Amplifiers Circuit Ideas, 1987, p. 13.

Chapter 2

Fig. 2-1. Analog Devices, Data Acquisition Databook, 1982, p. 4-56.

Fig. 2-2. Signetics, Analog Data Manual, 1982, p. 3-71.

Fig. 2-3. (TM) Siliconix, Siliconix Analog Switch and IC Product Data Book, 1/82, p. 7-29.

Chapter 3

Fig. 3-1. Precision Monolithics Inc., 1981 Full Line Catalog, p. 11-55.

Fig. 3-2. Reprinted with permission of Analog Devices, Inc., Data Acquisition Databook, 1982, p. 8-20.

Fig. 3-3. Electronic Engineering, 8/85, p. 30.

Fig. 3-4. Precision Monolithics Inc., 1981 Full Line Catalog, p. 11-55.

Fig. 3-5. Precision Monolithics Inc., 1981 Full Line Catalog, p. 16-10.

Fig. 3-6. Precision Monolithics Inc., 1981 Full Line Catalog, p. 11-54.

Fig. 3-7. Precision Monolithics Inc., 1981 Full Line Catalog, p. 16-159.

Fig. 3-8. Intersil, Applications Handbook, 1988, p. 2-38.

Fig. 3-9. Electronic Engineering, 11/86, p. 40.

Fig. 3-10. Intersil, Applications Handbook, 1988, p. 3-183.

Fig. 3-11. Ferranti, Technical handbook Vol. 10, Data Converters, 1983, p. 1-25.

Fig. 3-12. Courtesy of Motorola Inc., Linear Integrated Circuits, 1979, p. 4-50.

Fig. 3-13. (c) Siliconix, Siliconix Analog Switch and IC Product Data Book, 1/82, p. 8-5.

Fig. 3-14. (c) Siliconix, Siliconix Analog Switch and IC Product Data Book, 1/82, p. 8-4.

Fig. 3-15. Precision Monolithics Inc., 1981 Full Line Catalog, p. 6-10.

Fig. 3-16. (c) Siliconix, Siliconix Analog Switch and IC Product Data Book, 1/82, p. 8-5.

Fig. 3-17. Courtesy of Fairchild Camera and Instrument Corp., Linear Databook, 1982, p. 7-7.

Fig. 3-18. Courtesy of Motorola Inc., Linear Integrated Circuits, 1979, p. 3-17.

Fig. 3-19. Reprinted with permission of Analog Devices, Inc., Data Acquisition Databook, 1982, p. 10-50.

Fig. 3-20. Siliconix, Integrated Circuits Data Book, 3/85, p. 4-15.

Fig. 3-21. Precision Monolithics Inc., 1981 Full Line Catalog, p. 6-85.

Chapter 4

Fig. 4-1. Courtesy of Fairchild Camera and Instrument Corp., Linear Databook, 1982, p. 4-119.

Fig. 4-2. Courtesy of Fairchild Camera and Instrument Corp., Linear Databook, 1982, p. 4-178.

Fig. 4-3. Courtesy of Fairchild Camera and Instrument Corp., Linear Databook, 1982, p. 4-179.

Fig. 4-4. Courtesy of Fairchild Camera and Instrument Corp., Linear Databook, 1982, p. 4-119.

Fig. 4-5. Texas Instruments, Linear and Interface Circuits Applications, Vol. 1, 1985, p. 3-9.

Fig. 4-6. Texas Instruments, Linear and Interface Circuits Applications, Vol. 1, 1985, p. 3-7.

Fig. 4-7. Motorola, Linear Integrated Circuits, 1979, p. 3-147.

Fig. 4-8. Raytheon, Linear and Integrated Circuits, 1984, p. 6-205.

Fig. 4-9. Electronic Engineering, 2/47, p. 47.

Fig. 4-10. Courtesy, William Sheets.

Fig. 4-11. Reprinted with the permission of National Semiconductor Corp., Data Conversion/Acquisition Databook, 1980, p. 3-15.

Fig. 4-12. Reprinted with the permission of National Semiconductor Corp., Audio/Radio Handbook, 1989, p. 2-58.

Fig. 4-13. Courtesy of Motorola Inc., Linear Integrated Circuits, 1979, p. 3-131.

Fig. 4-14. Popular Electronics, Fact Card No. 59.

Fig. 4-15. Reprinted with the permission of National Semiconductor Corp., Audio/Radio Handbook, 1980, p. 2-59.

Fig. 4-16. Precision Monolithics Inc., 1981 Full Line Catalog, p. 16-116.

Fig. 4-17. Intersil, Component Data Catalog, 1987, p. 7-45.

Fig. 4-18. Signetics Analog Data Manual, 1982, p. 3-77.

Fig. 4-19. 73 Amateur Radio, 1/79, p. 127.

Fig. 4-20. Reprinted with permission of Analog Devices, Inc., Data Acquisition Databook, 1982, p. 4-97.

Fig. 4-21. 73 Amateur Radio, 4/79, p. 42.

Fig. 4-22. Popular Electronics, Fact Card No. 117.

Fig. 4-23. Reprinted with permission of Analog Devices, Inc., Data Acquisition Databook, 1982, p. 4-104.

Chapter 5

Fig. 5-1. Reprinted with the permission of National Semiconductor Corp., Linear Databook, 1982, p. 3-50.

Fig. 5-2. Hybrid Products Databook, 1982, p. 17-131.

Fig. 5-3. Reprinted with the permission of National Semiconductor Corp., Linear Databook, 1982, p. 3-157.

Fig. 5-4. Precision Monolithics Inc., 1981 Full Line Catalog, p. 7-11.

Fig. 5-5. Precision Monolithics Inc., 1981 Full Line Catalog, p. 16-158.

Fig. 5-6. Popular Electronics, Fact Card No. 117.

Fig. 5-7. Popular Electronics, Fact Card No. 117.

Chapter 6

Fig. 6-1. Linear Technology Corp., 1986 Linear Databook, p. 2-56.

Fig. 6-2. Siliconix, Integrated Circuits Data Book, 3/85, p. 10-62.

Fig. 6-3. Signetics Analog Data Manual, p. 75.

Fig. 6-4. Precision Monolithics Inc., 1981 Full Line Catalog, p. 16-115.

Fig. 6-5. Courtesy of Fairchild Camera and Instrument Corp., Linear Databook, 1982, p. 4-178.

Fig. 6-6. Precision Monolithics Inc., 1981 Full Line Catalog, p. 16-114.

Fig. 6-7. Harris Semiconductor, Linear and Data Acquisition Products, p. 2-85.

Fig. 6-8. Intersil, Component Data Catalog, 1987, p. 7-45.

Fig. 6-9. 303 Dynamic Electronic Circuits, TAB Book No. 1060, p. 289.

Chapter 7

Fig. 7-1. Harris, Analog Product Data Book, 1988, p. 10-16.

Fig. 7-2. Harris, Analog Product Data Book, 1988, p. 10-16.

Fig. 7-3. Electronic Engineering, 10/48, p. 45.

Fig. 7-4. Precision Monolithics Inc., 1981 Full Line Catalog, p. 16-116.

Fig. 7-5. 73 for Radio Amateurs, 2/86, p. 10.

Fig. 7-6. Siliconix, Integrated Circuits Data Book, 1988, p. 13-181.

Fig. 7-7. Reprinted from EDN, 9/29/88, (c) 1989 Cahners Publishing Co., a division of Reed Publishing USA.

Fig. 7-8. Reprinted with permission from Electronic Design. Copyright 1989, Penton Publishing.

Fig. 7-9. Intersil Data Book, 5/83, p. 3-135.

Fig. 7-10. Signetics Analog Data Manual, 1982, p. 4-8.

Chapter 8

Fig. 8-1. Reprinted with the permission of National Semiconductor Corp., Audio/Radio Handbook, 1980, p. 2-56.

Fig. 8-2. Reprinted with the permission of National Semiconductor Corp., Audio/Radio Handbook, 1980, p. 2-58.

Fig. 8-3. Popular Electronics, Fact Card No. 101.

Fig. 8-4. Popular Electronics, Fact Card No. 104.

Fig. 8-5. Reprinted from EDN, 3/16/89, (c) 1989 Cahners Publishing Co., a division of Reed Publishing USA.

Fig. 8-6. Electronics, 9/76, p. 100.

Chapter 9

Fig. 9-1. National Semiconductor Corp., Linear Applications Databook, p. 1096.

Fig. 9-2. EXAR, Telecommunications Databook, 1986, p. 7-24.

Fig. 9-3. EXAR, Telecommunications Databook, 1986, p. 7-24.

Fig. 9-4. Electronic Engineering, 12/84, p. 33.

Fig. 9-5. Electronic Engineering, 11/85, p. 31.

Fig. 9-6. Courtesy, William Sheets.

Fig. 9-7. Texas Instruments, Linear and Interface Circuits Applications, Vol. 1, 1985, p. 2-11.

Fig. 9-8. Courtesy, William Sheets.

Fig. 9-9. Courtesy, William Sheets.

Fig. 9-10. Courtesy, William Sheets.

Fig. 9-11. Reprinted with the permission of National Semiconductor Corp., Data Conversion/Acquisition Databook, 1980, p. 3-117.

Fig. 9-12. Reprinted from Electronics, 12/78, p. 124. Copyright 1978, McGraw Hill Inc., All rights reserved.

Fig. 9-13. National Semiconductor Corp., Linear Applications, p. 1083.

Fig. 9-14. Electronics Today International, 11/74, p. 67.

Fig. 9-15. Courtesy of Fairchild Camera and Instrument Corp., Linear Databook, 1982, p. 4-180.

Fig. 9-16. Texas Instruments, Linear and Interface Circuits Applications, 1985, Vol. 1, p. 3-10 and 3-11.

Fig. 9-17. Reprinted with the permission of National Semiconductor Corp., Hybrid Products Databook, 1982, 17-132.

Chapter 10

Fig. 10-1. Reprinted with the permission of National Semiconductor Corp., Data Conversion/Acquisition Databook, 1980, p. 3-23.

Fig. 10-2. Raytheon, Linear and Integrated Circuits, 1989, p. 4-189.

Fig. 10-3. (c) Siliconix, Siliconix Analog Switch and IC Product Data Book, 1/82, p. 6-9.

Fig. 10-4. Electronics Today International, 10/78, p. 26.

Fig. 10-5. Harris Semiconductor, Linear and Data Acquisition Products, p. 2-84.

Fig. 10-6. Courtesy of Motorola Inc., Motorola Semiconductor Library Vol. 6, Series B, p. 3-126.

Fig. 10-7. Intersil, Component Data Catalog, 1987, p. 8-102.

Fig. 10-8. Ham Radio, 2/78, p. 72.

Fig. 10-9. Signetics Analog Data Manual, p. 401.

Fig. 10-10. Precision Monolithics Inc., 1981 Full Line Catalog, p. 6-58.

Fig. 10-11. Popular Electronics, Fact Card No. 59.

Chapter 11

Fig. 11-1. Ham Radio, 7/76, p. 69.

Fig. 11-2. Popular Electronics, 10/89, p. 42.

Fig. 11-3. Ham Radio, 5/89, p. 26.

Fig. 11-4. Radio Electronics, 9/89, p. 47.

Fig. 11-5. 73 Amateur Radio.

Fig. 11-6. National Semiconductor Corp., Transistor Databook, 1982, p. 7-27.

Chapter 12

Fig. 12-1. National Semiconductor Corp., 1984 Linear Supplement Databook, p. S5-143.

Fig. 12-2. Teledyne Semiconductor Publication DG-114-87, p. 7.

Fig. 12-3. (c) Siliconix Inc., Analog Switch and IC Product Data Book, 1/82, p. 7-30.

Fig. 12-4. Reprinted with the permission of National Semiconductor Corp., Linear Databook, 1982, p. 9-140.

Fig. 12-5. Reprinted with the permission of Analog Devices, Inc., Data Acquisition Databook, 1982, p. 12-20.

Fig. 12-6. Reprinted with the permission of National Semiconductor Corp., Linear Databook, 1982, p. 9-143.

Fig. 12-7. Reprinted from EDN, 3/21/85, (c) 1989 Cahners Publishing Co., a division of Reed Publishing USA.

Chapter 13

Fig. 13-1. Reprinted with the permission of National Semiconductor Corp., Linear Applications Handbook, 1982, p. AN240-5.

Fig. 13-2. Electronic Engineering, Applied Ideas, 11/88, p. 28.

Fig. 13-3. Reprinted from EDN, 10/2/86, (c) 1989 Cahners Publishing Co., a division of Reed Publishing USA.

Fig. 13-4. Linear Technology Corp., Linear Databook, 1986, p. 8-13.

Fig. 13-5. Linear Technology Corp., Linear Databook, 1986, p. 8-43.

Fig. 13-6. Linear Technology Corp., Linear Databook, 1986, p. 2-112.

Fig. 13-7. Reprinted from EDN, 10/1/87, (c) 1989 Cahners Publishing Co., a division of Reed Publishing USA.

Fig. 13-8. Signetics, 1987 Linear Data Manual Vol. 1: Communications, 2/87, p. 4-311.

Fig. 13-9. General Electric/RCA, BiMOS Operational Amplifiers Circuit Ideas, 1987, p. 20.

Fig. 13-10. General Electric/RCA, BiMOS Operational Amplifiers Circuit Ideas, 1987, p. 11.

Fig. 13-11. Siliconix, MOSpower Applications Handbook, p. 6-178.

Fig. 13-12. Electronics Today International, 10/77, p. 45.

Fig. 13-13. (c) Siliconix, Siliconix Analog Switch and IC Product Data Book, 1/82, p. 7-29.

Fig. 13-14. Reprinted with the permission of National Semiconductor Corp., National Semiconductor CMOS Databook, 1981, p. 3-61.

Fig. 13-15. (c) Siliconix, T100/T300 Applications.

Fig. 13-16. Reprinted with the permission of National Semiconductor Corp., *Linear Applications Handbook*, 1982, p. AN240-2.

Fig. 13-17. Reprinted with permission of Analog Devices, Inc., *Data Acquisition Databook*, 1982, p. 6-27.

Fig. 13-18. Reprinted with the permission of National Semiconductor Corp., *Linear Applications Handbook*, 1982, p. 8-258.

Fig. 13-19. (c) Siliconix, *Siliconix Analog Switch and IC Product Data Book*, 1/82, p. 7-31.

Fig. 13-20. Signetics, *Analog Data Manual*, 1982, p. 3-15.

Fig. 13-21. RCA Corporation, Solid State Division, Digital Integrated Circuits Application Note ICAN-6346, p. 4.

Fig. 13-22. (c) Siliconix, *MOSPOWER Design Catalog*, 1/83, p. 6-42.

Fig. 13-23. Signetics, *Analog Data Manual*, 1982, p. 8-14.

Fig. 13-24. Reprinted with permission of Analog Devices, Inc., *Data Acquisition Databook*, 1982, p. 4-56.

Fig. 13-25. Motorola, *Motorola TMOS Power FET Design Ideas*, p. 35.

Fig. 13-26. Reprinted from EDN, 8/17/89, (c) Cahners Publishing Co., a division of Reed Publishing, USA.

Fig. 13-27. Reprinted from EDN, 3/3/88, (c) Cahners Publishing Co., a division of Reed Publishing, USA.

Fig. 13-28. Reprinted with the permission of National Semiconductor Corp., *Linear Applications Handbook*, 1982, p. 3-50.

Fig. 13-29. Electronic Engineering, 12/77, p. 19.

Fig. 13-30. Reprinted from EDN, 3/7/85, (c) Cahners Publishing Co., a division of Reed Publishing, USA.

Fig. 13-31. Linear Technology Corp., *Linear Databook*, 1986, p. 5-17.

Fig. 13-32. Precision Monolithics Incorporated 1981 Full Line Catalog. p. 16-142.

Fig. 13-33. Signetics, *1987 Linear Data Manual Vol. 2: Industrial*, 2/87, p. 7-62.

Fig. 13-34. (c) Siliconix, *Siliconix Analog Switch and IC Product Data Book*, 1/82, p. 7-30.

Fig. 13-35. Signetics, *Analog Data Manual*, 1982, p. 6-20.

Fig. 13-36. Electronics Today International, 10/77, p. 39.

Chapter 14

Fig. 14-1. Signetics, *Analog Data Manual*, 1982, p. 6-21.

Fig. 14-2. Signetics, *Analog Data Manual*, 1982, p. 6-21.

Fig. 14-3. Motorola, *Motorola TMOS Power FET Design Ideas*, 1985, p. 39.

Fig. 14-4. Maxim, 1986 Power Supply circuits, p. 44.

Fig. 14-5. Radio-Electronics, 4/85, p. 80.

Fig. 14-6. Linear Technology Corp., *Linear Applications Handbook*, 1987, p. AN8-9.

Fig. 14-7. (c) Siliconix, *Siliconix Analog Switch and IC Product Data Book*, 1/82, p. 6-15.

Fig. 14-8. (c) Siliconix, *MOSPOWER Design Catalog*, 1/83, p. 6-41.

Fig. 14-9. Signetics, *Analog Data Manual*, 1982, p. 6-13.

Fig. 14-10. Electronics Today International, 8/79, p. 99.

Fig. 14-11. Motorola, *Motorola TMOS Power FET Design Ideas*, 1985, p. 41.

Fig. 14-12. Motorola, *Linear Integrated Circuits*, p. 5-145

Fig. 14-13. Electronic Engineering, 11/76, p. 23.

Fig. 14-14. Datel, *Data Conversion Components*, p. 6-18.

Fig. 14-15. Electronics Today International, 9/75, p. 65.

Fig. 14-16. R-E Experimenters Handbook, 1987, p. 129.

Fig. 14-17. Popular Electronics, 6/89, p. 25.

Fig. 14-18. Maxim, *Seminar Applications Book*, 1988/89, p. 81.

Fig. 14-19. Motorola, *Motorola TMOS Power FET Design Ideas*, 1985, p. 36.

Fig. 14-20. Electronic Engineering, 8/83, p. 141.

Fig. 14-21. Signetics, *1987 Linear Data Manual Vol. 2: Industrial*, 2/87, p. 5-368.

Fig. 14-22. Siliconix, *Integrated Circuits Data Book*, 3/85, p. 5-17.

Fig. 14-23. Signetics, *1987 Linear Data Manual Vol. 2: Industrial*, 2/87, p. 7-65.

Fig. 14-24. Maxim, *Seminar Applications Book*, 1988/89, p. 149.

Fig. 14-25. Maxim, *Seminar Applications Book*, 1988/89, p. 83.

Fig. 14-26. Reprinted from EDN, 3/3/88, (c) 1989 Cahners Publishing Co., a division of Reed Publishing USA.

Fig. 14-27. Maxim, Seminar Applications Book, 1988/89, p. 78.

Fig. 14-28. Reprinted with the permission of National Semiconductor Corp., Audio/Radio Handbook, 1980, p. 4-28.

Chapter 15

Fig. 15-1. Electronics Today International, 4/81, p. 86.

Fig. 15-2. Linear Technology Corp., Linear Applications Handbook, 1987, p. AN7-2.

Fig. 15-3. Reprinted with the permission of National Semiconductor Corp., Linear Databook, 1982, p. 8-258.

Fig. 15-4. Electronics Today International, 10/78, p. 101.

Chapter 16

Fig. 16-1. Signetics, Analog Data Manual, 1983, p. 10-99.

Fig. 16-2. Signetics, 1987 Linear Data Manual Vol. 2: Industrial, 11/86, p. 4-136.

Chapter 17

Fig. 17-1. Electronic Engineering, 12/75, p. 11.

Fig. 17-2. Electronic Engineering, 7/76, p. 23.

Fig. 17-3. Reprinted with the permission of National Semiconductor Corp., Linear Applications Handbook, 1982, p. D-7.

Fig. 17-4. Reprinted with permission of Analog Devices, Inc., Data Acquisition Databook, 1982, p. 12-20.

Fig. 17-5. Reprinted with permission from Electronic Design. Copyright 1975, Penton Publishing.

Fig. 17-6. Linear Technology Corp., Linear Applications Handbook, 1987, p. AN9-13.

Fig. 17-7. Teledyne Semiconductor, Data Acquisition IC Handbook, 1985, p. 9-7.

Fig. 17-8. Signetics, 1987 Linear Data Manual Vol. 2: Industrial, 2/87, p. 7-63.

Fig. 17-9. (c) Siliconix, Analog Switch and IC Product Data Book, 1/82, p. 1-25.

Fig. 17-10. Courtesy of Fairchild Camera and Instrument Corp., Linear Databook, 1982, p. 7-7.

Fig. 17-11. Reprinted with permission of Analog Devices, Inc., Data Acquisition Databook, 1982, p. 12-19.

Fig. 17-12. National Semiconductor Corp., 1984 Linear Supplemental Databook, p. S5-142.

Fig. 17-13. Linear Technology Corp., Linear Applications Handbook, 1987, p. AN9-14.

Fig. 17-14. Analog Devices, Data Acquisition Databook, 1982, p. 6-27.

Fig. 17-15. Gernsback Publications Inc., 44 New Ideas, 1985, p. 8.

Fig. 17-16. Reprinted with permission from Electronic Design. Copyright 1967, Penton Publishing.

Index